Groundwater and Society:

Resources, Tensions and Opportunities

NOTE

The designations employed and the presentation of the material in this publication do not imply the expression of any opinion whatsoever on the part of the Secretariat of the United Nations concerning the legal status of any country, territory, city or area or of its authorities, or concerning the delimitation of its frontiers or boundaries.

The term "country" as used in the text of this publication also refers, as appropriate, to territories or areas.

ST/ESA/265

United Nations publication
Sales No. E.99.II.A.1
ISBN 92-1-104485-5

Litho in United Nations, New York
98-35194—October 2000—2,810

Groundwater and Society:
Resources, Tensions and Opportunities

Themes in groundwater management
for the twenty-first century

Jacob J. Burke, DESA
Marcus H. Moench, ISET

DESA
United Nations Department of Economic and Social Affairs
with the support of the Economic Commmission for
Latin America and the Caribbean (ECLAC)

ISET
Institute for Social and Environmental Transition

Foreword

by Nitin Desai, Under-Secretary-General,
United Nations Department of Economic and Social Affairs (DESA)

In 1960, the United Nations published *Large-scale Groundwater Development* (United Nations, 1960) in response to a specific request from the Economic and Social Council of the United Nations (ECOSOC). The book was written by a set of eminent hydrogeologists against a background of intensive development in irrigated agriculture accompanied by borehole mechanization, and the beginnings of urban expansion in response to population growth. It was full of clear advice on groundwater development, recognizing social and economic values and presenting well-thought-out practical steps and principles that would lead to sustainable development of groundwater resources against these changing consumption patterns.

Arguably, had the recommendations of that book been implemented there would be no need for the present volume. Since 1960, the UN system supported many urgently needed interventions in groundwater development, data collection, analysis and training. But where are we now—what was missing or unforeseen? Do we really know what we have done to our groundwater resources and are we fully aware of the impact this has had on those who are utterly dependent upon them for sources of freshwater and livelihoods? This book attempts to take stock of the development and abuse of groundwater that have taken place during this century and to measure present approaches to groundwater management against the reality of declining water-tables and polluted aquifers. More importantly, it goes on to discuss the impact these have had on people, their livelihoods, communities and environments. The prospects for sustainable development are then examined in this sobering light. While we may have inflicted irreparable damage on many of our prime aquifers, there appears to be scope for damage control and sustainable development if we approach groundwater management with a much more sharpened appreciation of how people organize themselves around this critical resource.

Acknowledgements

This book is based on the findings of an ad hoc meeting of experts held at the United Nations from 10 to 12 December 1997 and organized jointly by the United Nations Department of Economic and Social Affairs (DESA) and the Institute for Social and Environmental Transition. Participants in the ad hoc meeting of experts included: Charles Abdalla, Pennsylvania State University (United States); Ian Acworth, University of New South Wales (Australia); David Brooks, International Development Research Centre (Canada); Jacob Burke, Technical Adviser, DESA; Hector Garduno, Comisión Nacional del Agua (Mexico); Uri Golani, a former UN Interregional Adviser (Israel); Brian Morris, British Geological Survey (United Kingdom); Marcus Moench, ISET (United States); Anthony Navoy, U.S. Geological Survey (United States); Dominique Poitrinal, Centre thématique d'eau, Bureau de recherches géologiques et minières (France); Jefter Kuziwa Sakupwanya, Ministry of Rural Water Resources and Water Development (Zimbabwe); Claude Sauveplane, Interregional Adviser, DESA; and Tushaar Shah, ISET (India). The principal authors of the report are Jacob Burke of DESA and Marcus Moench of ISET, with technical input from Claude Sauveplane of DESA, Miguel Solanes of ECLAC and David Kromm of Kansas State University. Observations and comments on the first draft, along with additional input, in particular materials for boxes and diagrams, were provided by all participants. The authors are particularly grateful to Marcia Brewster of DESA for her extensive editorial review and to Fabiola Knight of DESA for keeping things on track.

Cover photo: Extract of Landsat image, central Sudan

Contents

Executive summary

This book concerns itself with the specific management of groundwater as a component of integrated water management. It is predicated on the observation that groundwater is an often unnoticed and unacknowledged cornerstone in the foundation of many economic and environmental systems. Even in the contemporary accounts of "integrated" water management, the special character of the aquifer systems that underpin the resource base is rarely discussed. It comes as no surprise that in many regions of the world, the groundwater resource base and the social, economic and environmental systems dependent on it are under threat from over-abstraction and pollution. It is important to recognize that this has been a recent phenomenon. The scale and intensity of the abstraction and pollution have been apparent only in the past 20 years. Prior to this, groundwater was seen as a ubiquitous and reliable source of high-quality water. What is equally apparent is that the evolution of effective management systems to address these threats will be a long-term process requiring both sustained political commitment and improvements in basic data and scientific understanding. These are essential for interpreting not only complex groundwater resource dynamics but also the patterns of socio-economic demands that are placed upon these resources. Waiting for data should not, however, become an excuse to delay action. *Simple indicators, such as major land-use changes, long-term water-level declines and increases in salinity, pathogens or key pollutants, signal the need for management. In many cases immediate action to reduce pumping, control pollution sources and collect more detailed data on aquifer dynamics is essential to avoid irreversible economic, social and environmental damage.* Management initiatives of this type—even if only partially successful—can provide a critical breathing-space while more comprehensive and integrated management systems based on better understanding of groundwater dynamics and the socio-economic systems that exploit them are developed.

The social, economic and environmental values associated with groundwater are often unrecognized and undervalued. Groundwater is the most reliable source of supply for potable water and supports a wide array of economic and environmental services. Of these, agriculture, the largest abstractor of groundwater, is less sensitive to water quality but is generally the highest-volume user. The role of groundwater in agriculture is important to recognize. Groundwater is the primary buffer against drought, and areas with access to groundwater irrigation are generally able to achieve higher agricultural yields. If climatic variability increases, as many analysts predict will be the case with global climatic change, the buffering value of groundwater will be a particularly important factor determining society's ability to meet basic food security, drinking-water supply and environmental needs that depend on reliable water sources. Even without climatic change, supporting global populations will require reliable water supplies. At the start of the twenty-first century, over 50 per cent of the world's population will reside in urban areas—a dramatic increase from the 30 per cent in urban population in 1950. Most of this urban population growth will occur in developing countries. Of the 23 mega-cities estimated to have populations of over 10 million by the year 2000, 12 are heavily dependent on groundwater and, with the exception of London, all are in the developing world.[1] Furthermore, other mega-cities, such as Los Angeles, see groundwater as a fundamental component of their water-supply planning or, as in the case of New Delhi, have large populations that are not served by the municipal system and that rely on groundwater (which is often polluted) as their primary source of supply. Groundwater is thus central to meeting the large-scale needs for food security and urban drinking-water supply.

Equally important to its role as a critical source of water supply for agricultural and municipal uses, groundwater plays a more subtle role related to poverty alleviation, health and social vulnerability. Access to groundwater is perhaps the most critical factor enabling many rural populations to maintain sustainable livelihoods. Assured water supplies greatly reduce the risks poor farmers face when investing in such agricultural inputs as seed and fertilizer. Secure water supplies enable them to increase yields, income levels, savings and capital formation substantially. Similar effects occur where health is concerned. Groundwater is generally less vulnerable to pollution than surface-water sources. It is also often available in close proximity to points of use. In combination, these factors reduce the risks from water-borne disease and reduce time spent in collecting water from distant locations. Fewer sick days and reductions in time "wasted" collecting water translate into more time available for more productive purposes. Overall, by enabling individuals to accumulate reserves, access to groundwater enables rural populations to reduce their vulnerability, not just to drought, but to the full range of natural, economic and social hazards that generate much rural poverty.

Finally, groundwater plays a crucial role in the environment and water-related ecosystems. It is the primary source of base flow in streams and rivers and is a major water source for most surface vegetation communities and wetlands. But the aquifers that host the groundwater are also the ultimate terrestrial "sink" for land-based sources of pollution and accumulate waste products over time. *Man-induced changes in groundwater conditions (circulation and aquifer status/configuration) often result in complex environmental and socio-economic impacts that are difficult to predict or remediate.*

As a consequence, the positive role of groundwater as a cornerstone in the foundation of regional socio-economic and environmental systems is severely compromised. In many cases, the productive status of large groundwater bodies has been lost because of over-abstraction and pollution which, under current recharge regimes and pollution intensity, will not be regained. Recent decades have seen a global explosion in both installed pumping capacity and potential sources of

[1] Mexico City, Tehran, Shanghai, Buenos Aires, Jakarta, Karachi, Dhaka, Manila, Cairo, Bangkok, London and Beijing.

pollution. In many regions these are having a dramatic effect on groundwater availability. Piezometric levels in some arid regions are declining at rates greater than 3 metres a year and groundwater is being extracted at unsustainable rates. Pollution and quality deterioration are also reducing the availability of usable groundwater supplies. Once aquifers are polluted, their remediation is often economically or technically impossible.

The absolute availability of groundwater resources is, however, far less important for the sustainability of economic and environmental uses than characteristics of hydrologic systems dependent on specific groundwater conditions. *Pollution and declines in water level and water quality frequently affect the sustainability of groundwater-dependent uses, whether or not the resource base itself is threatened with physical exhaustion or severe degradation.* It matters little if a basin contains 6,000 metres of sediments saturated with high-quality groundwater (as the Gangetic Basin does) if groundwater levels have declined to the point where extraction is uneconomic, the shallow wells of poor farmers run dry and dry-season base flows in rivers and lakes decline. Disruption of hydrogeologic systems can easily threaten food security, drinking-water supplies and the environment or cause massive poverty even when the amount of water physically available remains comparatively large. This crucial point is often missed in debates over the importance of groundwater management.

Because the links between users and the resource are often not apparent, and because many benefits associated with groundwater are public goods, the overall economic value of groundwater goes unrecognized. As a consequence, groundwater resources tend to be used with little regard for economic and other externalities. The direct user may be forced to internalize some impacts (of water-level decline, for example), but the common-pool nature of the resource and the fact that many of the services it provides serve a public interest (such as environmental maintenance, health and poverty alleviation) create many challenges for management. From an economic perspective, the total value of groundwater is the sum of extractive and *in situ* values. Extractive values tend to accrue to individual users while *in situ*, or existence, values are generally public goods. This structural dichotomy creates strong social incentives for over-extraction. It also leads to competition between individuals, sectors, regions and, in some cases, nations. There is often a "race to the pumphouse" as individuals and different user groups seek to capture for themselves as much as possible of the benefits associated with the extractive value of the resource.

Beyond valuation and competition issues, the distribution of groundwater resources and use creates major challenges for management. Extraction points are generally dispersed and groundwater conditions often depend on the decisions of numerous individual landowners and well owners. Continuous aquifer systems, however, can extend across multiple geographic, administrative and political regions. In this situation, individual users generally have little hope of influencing groundwater conditions through isolated actions under their direct control.

In this overall context, some form of regulatory management (whether imposed by the government or user-based organizations or through informal social mechanisms) is essential to maintain the manifest public interest and existence values of the resource. The role regulation can play is, however, specifically conditioned by socio-economic circumstances and the configuration and status of the aquifers being used. In many developing-country situations, individual users number in the hundreds of thousands (in the case of India, there are over 20 million private wells). Direct regulation of individual users by government is often politically and practically impossible. Furthermore, management will need to vary at local to regional scales in order to reflect both social and hydrogeological factors. This can make centralized regulation inefficient and ineffective. In many cases, much more adaptive approaches to local resource management involving a high level of stakeholder[2] participation are essential. Wherever possible, these should be *effectively coordinated at the respective river basin and aquifer scales.* It is important to recognize, however, that many types of management (such as the control of pollution from concentrated urban or industrial sources or management to address localized pumping depressions) can be effective even without coordination at a larger scale. The absence of an effective mechanism for coordination should not be an excuse for the lack of management at local levels.

In this respect, it is clear that no action would be implementable or enforceable without general legislation *declaring a public interest in the resource* and enabling the government or other public or private bodies to undertake and enforce appropriate regulation.

Effective approaches to groundwater management cannot generally depend on single mechanisms. Political support for management rests on public and policy-maker understanding of management needs. Education is essential. Economic incentives that reflect the full value of groundwater resources are also important. These need to work in concert with legal and regulatory frameworks that both enable local populations to develop management approaches suited to regional conditions and provide avenues for higher-level interventions to address the actions of large individual consumers or polluters. Evolution of effective management approaches thus requires a delicate balancing act among individual, community and government roles and educational, economic, legal and regulatory mechanisms. Furthermore, because groundwater issues and the technical, social, economic, cultural and political factors influencing management options vary greatly between regions, no single template for management can be developed. As a result, *the development of effective approaches will, in most cases, require a long-term process through which viable national, regional and local systems can evolve.*

Although the evolution of effective groundwater management systems is not an easy process, it should not be an excuse to delay action. Management restrictions, such as reductions in pumping or controls on waste disposal, can be relaxed if, at a later stage, more detailed information and understanding of aquifer dynamics suggest they are unnecessary. In general, the risks inherent in delaying management outweigh the potential costs associated with initiating management on the basis of imperfect information. Management restrictions, if later proved unnecessary, may have resulted in a *temporary* loss of economic and social benefits. If, however, management action is needed but not initiated, *irreversible* damage to the groundwater resource base may occur and result in far greater losses.

In a larger sense, the complexity of management makes the collection of targeted high-quality information on groundwater systems and their dependent communities essential. It requires both long-term efforts to collect basic data and

[2] Throughout this book, the authors use this term in its colloquial sense (and not strict literal sense) as a matter of convenience. Stakeholders in groundwater management are taken to mean all users, institutions and those affected by use and management who collectively or individually hold a "stake" in the resource and the way it is managed.

research to constantly refine understanding of both groundwater and socio-economic processes. Groundwater research and socio-economic research need to be coupled. Hydrogeological research on its own is of relatively little use for management purposes unless it is complemented by information that links groundwater problems with their social and economic context. Furthermore, unless information is used to inform and educate the general public and policy makers, it will be difficult to develop a social consensus regarding problems and potential solutions. For this reason, information also has to be clearly presented to inform and educate the general public and policy makers. Without a high degree of social consensus, management systems will often be politically impossible to implement.

Developing the social consensus necessary for effective groundwater management is further complicated by the deeply entrenched nature of existing organizations and by culturally defined ethical positions. In many countries, governmental water resource agencies have been designed according to centralized managerial principles and emphasize surface-water resources. As a result, they tend to rely on technocratic options to address the problems they perceive as most important. This can bring them into direct conflict with established users and local communities, who often have different perceptions regarding the nature of problems. Management debates, as a result, often become polarized and deadlocked over differing perceptions and locus of control issues. Therefore, the process of policy-making and policy implementation requires that governments, and their technicians, conduct extensive awareness-building campaigns to educate stakeholders, to demonstrate the losses associated with inaction, and to interest and involve users and local communities in management efforts. Ethical issues also complicate management. In some societies, groundwater is or has been linked to landownership, while in others it is viewed as a "common heritage" (not to be confused with the British "common law" system) to which all should have equal access, at least for basic needs. These conflicting positions are enshrined in religious doctrines such as the shariah (where the "right to thirst" is a basic principle) and in Western legal concepts rooted as early as Roman times (in which groundwater ownership follows landownership). At the same time, rights of access to groundwater have generally been linked to landownership. There is thus often an unclear distinction between the "private" nature of groundwater rights and "public" ownership of the re-source itself. This contradiction is brought to the fore by increased recognition of the need for water to be used more efficiently. Market mechanisms can play a major role in achieving efficiency objectives, and more emphasis is now being given to the nature of water as an economic resource in global debates. This emphasis translates into initiatives to clarify water rights, encourage water markets and issue "concessions" in some countries. However, the process may be still-born if there is no recognition that water resources, by their very "public" nature, require regulation and are not amenable to absolute free-market solutions. As these initiatives increase, tensions related to underlying ethical issues can also be expected to increase. If groundwater is viewed as a common heritage to which all have fundamental rights of access, the ethical basis of unregulated concessions or markets that allocate water depending on ability to pay becomes controversial. No conflict exists between the basic goal of efficient and effective management and the common heritage or private nature of groundwater resources. How efficient management is to be achieved, the sets of social objectives against which efficiency is measured (including maintenance of supplies for the poor and the environment), what sections of society play a role in decision-making and what management mechanisms are used all touch deep ethical roots and can become points of tension. This type of issue—as with many issues in society—may have no fundamentally correct answer. Such debates are, however, central to developing the social consensus on which political decisions regarding groundwater depend. This is why water law systems do not tend to adhere to philosophical neatness but devise pragmatic balances between the investment needs related to the economic features of water and the need for public regulation associated with the social and environmental dimensions of the resource.

In conclusion, social, economic, cultural and ethical dimensions are likely to be as important as technical dimensions in the evolution of approaches to address existing and emerging groundwater problems and ensure the sustainability of key social, economic and environmental systems. Basic data and information on aquifer condition and the projected demands, as well as effective user participation based on education and constructive dialogue, are essential inputs for resolving such complex issues. Over the long term, education of children is particularly important in order to shift society's awareness and understanding of both groundwater problems and management opportunities. Groundwater problems are generally not amenable to rapid solutions. Rather, they signal the need to develop management systems capable of flexibly addressing and predicting constraints as they arise. In many developing countries this process is in its infancy, and it is crucial to initiate and encourage whatever management actions are currently feasible while starting the long-term process of developing flexible, integrated management systems.

1 Introduction

Groundwater is, for many of us, an invisible resource. Because it is widely distributed in a range of water-producing geological structures (aquifers) and since it is much less dependent on recent precipitation than surface sources, it can provide a uniquely reliable source of high-quality water for human uses. In many respects, the differentiation between surface water and groundwater in this book is deliberate and fundamental. While the links between the two sources of water may be intimate, the management of groundwater presents a separate and, arguably, more complex set of issues that relate to the spatial and temporal dimensions of its occurrence and the way societies and their economies organize themselves around its use. Hence, this book concerns itself with groundwater management per se as a component of integrated water resource management. The broader aspects of integrated water management are well discussed elsewhere and for this the reader is directed towards accounts given by others (e.g., Mitchell, 1990; Winpenny, 1994). But to highlight the argument, even in these accounts, the discussion of groundwater management is conspicuous by its absence. Furthermore, groundwater has a number of special characteristics that are distinct from those addressed in general discussions of integrated water management. The advent and rapid spread of energized pumping technologies have enabled rapid groundwater development and the emergence of socio-economic systems dependent on its reliability. The development of groundwater continues to be predicated upon the perceived advantages of groundwater and the general assumption that the resource, as with surface-water systems, will be replenishable or the reserves so great that one generation's impact will be insignificant. Development has, however, occurred without an adequate understanding of the complex and vulnerable nature of groundwater systems. Aquifer characteristics and groundwater flow properties vary laterally, vertically and temporally. These combine to create dynamic, interdependent systems that can be disrupted in unpredictable ways as a result of rapid development. What is remarkable is the rate at which aquifer systems have been depleted and degraded through over-abstraction and pollution. The bulk of the damage has occurred in the last 50 years. Unlike surface-water systems, much of this loss is irreversible and therefore much more critical. More importantly, development has occurred with little appreciation of how societies and economies organize themselves to take advantage of the opportunities groundwater presents and to respond to management needs as they emerge. Understanding the socio-political economy (and culture) surrounding groundwater, and the management

Box 1 Participation, philosophy and pragmatism

Debates over the relative roles of government institutions and agencies and other stakeholders underlie differing approaches to groundwater management, as they do approaches to the management of most naturalresources. Many participants in management debates advocate broad-based "participatory" approaches due to philosophical considerations related to governance. From this perspective, broad-based participation by all stakeholders is viewed as an essential mechanism to counterbalance special interests and governmental excess. Here, stakeholders in groundwater management include all users, institutions and those affected by use and management who collectively or individually hold a "stake" in the resource and the way it is managed. The emphasis here is on the

"broad-based" participation by all stakeholders, including the government. Equally, effective participation can prompt and gain support for action in areas, such as environmental protection, where institutions are weak or indifferent. As such, it represents a cornerstone of democratic governance philosophies.

Whatever the governance philosophy, however, stakeholder participation is particularly important in the groundwater context. Unlike surface irrigation or drinking-water-supply systems, the structure of groundwater resource systems is poorly suited to centralized control. Abstraction is geographically dispersed and generally under the effective control of numerous individual well owners. There are few, if any, centralized points where flows within aquifer systems can be controlled. Furthermore, in many (if not most) situations, options for enhancing groundwater availability through recharge are limited. Management, as a result, depends on changes in abstraction and use. This is, in turn, dominantly a function of use decisions by individual well owners. From a purely pragmatic perspective, options for centralized governmental organizations to influence or control those individual use decisions tend to be extremely limited unless the support and collaboration of users can be obtained. This practical problem has, for example, been the core factor blocking adoption of legislation enabling groundwater regulation in India for more than two decades. Model legislation was written in 1972 but has yet to be widely adopted because of political opposition by users and because state governments and water resource agencies view the proposed regulatory approach as impossible to implement, given the millions of wells already existing in private hands.

These pragmatic considerations imply that stakeholder participation is likely to be essential in most cases regardless of the urgency of management needs or the political philosophy of the government. Stakeholder understanding of emerging problems and support for management should increase the scope of possible interventions. Participation around the development of wells has been a significant feature in some parts of India and Nepal. But it is essential to recognize that stakeholder participation is no panacea and may result in management approaches that are, from technical and other perspectives, suboptimal. As Kemper (1996) and Wilson (1998) have documented, stakeholders are often motivated by self-interest and advocate solutions that are technically or socially problematic. As a result, leaving governance philosophies aside, stakeholder participation represents a critical starting point for management but not one that should exclude other regulatory, economic or technical avenues.

The access to groundwater in the arid wadis of south Yemen has transformed landscapes and made settlement possible.

tools that may apply, is as important as understanding the physical systems. There are very few published accounts of this important aspect. Shah (1993) and Barraqué (1996, 1997a, 1997b) discuss issues in relation to India and Europe, but the issues raised apply globally. This form of engagement with the resource base is special to groundwater and raises the fundamental issue of participation in groundwater management by those who have an interest in the resource—"stakeholders", for want of a better term. The authors use the term "stakeholders" in its colloquial sense (and not strict literal sense) as a matter of convenience. Stakeholders in groundwater management are taken to mean all users, institutions and those affected by use and management who collectively or individually hold a "stake" in the resource and the way it is managed (see box 1).

Developed and developing countries alike are attempting to frame approaches to the sustainable development of their sovereign and shared water resources. In many cases the socio-economic and hydro-environmental imperatives do not leave much room for manoeuvre. This is particularly true for groundwater, where expansion of agricultural production and, to a lesser extent, domestic water-supply systems have induced unsustainable patterns of groundwater use. This document argues that social, institutional and political factors are the primary obstacles to sustainable management of the world's groundwater resources. Technologies for groundwater exploration and extraction are developing all the time, together with sophisticated techniques for understanding flow dynamics within complex aquifer systems.[3] Equally, hydrogeological tools are being added to the repertoire of conjunctive use and water conservation (Pyne, 1995; van der Merwe, 1998). Understanding of the social, economic, institutional and political dimensions essential for effective management, however, lags far behind.

1.1 Patterns of groundwater use—issues for sustainability

A large proportion of the world's population relies on groundwater as a primary source of supply for drinking and domestic uses. Indirectly and often unknowingly, their survival depends on it as a key input for sustainable food production. Agricultural economies in arid, semi-arid and humid zones have expanded through exploitation of the groundwater resource base. Reliance shifts to dependency in drought years when groundwater plays its critical role in buffering water-supply availability. The regular replenishment of aquifers through direct recharge processes from rainfall and indirect recharge from surface-water bodies is vital in sustaining shallow groundwater systems and maintaining the piezometric heads of deeper flow systems. This input is highly variable, particularly in arid and semi-arid zones where there is greater dependence upon groundwater.

Irrigated agriculture on Batinah coastal plain, Oman, is entirely dependent upon actively recharging groundwater systems—as illustrated in this Landsat image.

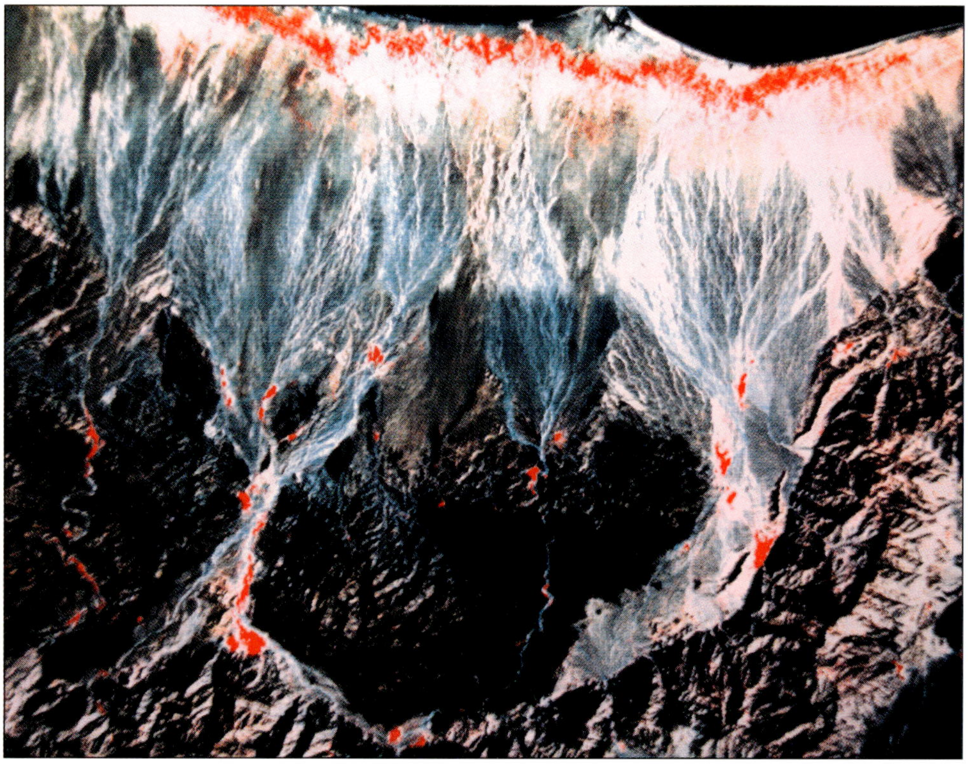

Source: Wimpey Laboratories

[3] Current scientific research on groundwater, which now includes physical and analytic tools from the petroleum industry, is advancing knowledge regarding the physical dynamics of groundwater systems. The fundamental issues of aquifer heterogeneity are well appreciated by the authors through both practical experience and state-of-the-art reviews of groundwater research and development. The contributions of geologists, geophysicists and sedimentologists are of fundamental importance to understanding aquifer parameters and identifying management options. The process by which this improved understanding can translate into practical groundwater management practice is one of the central concerns of this document.

Figure 1 Growth of energized pumpsets in India

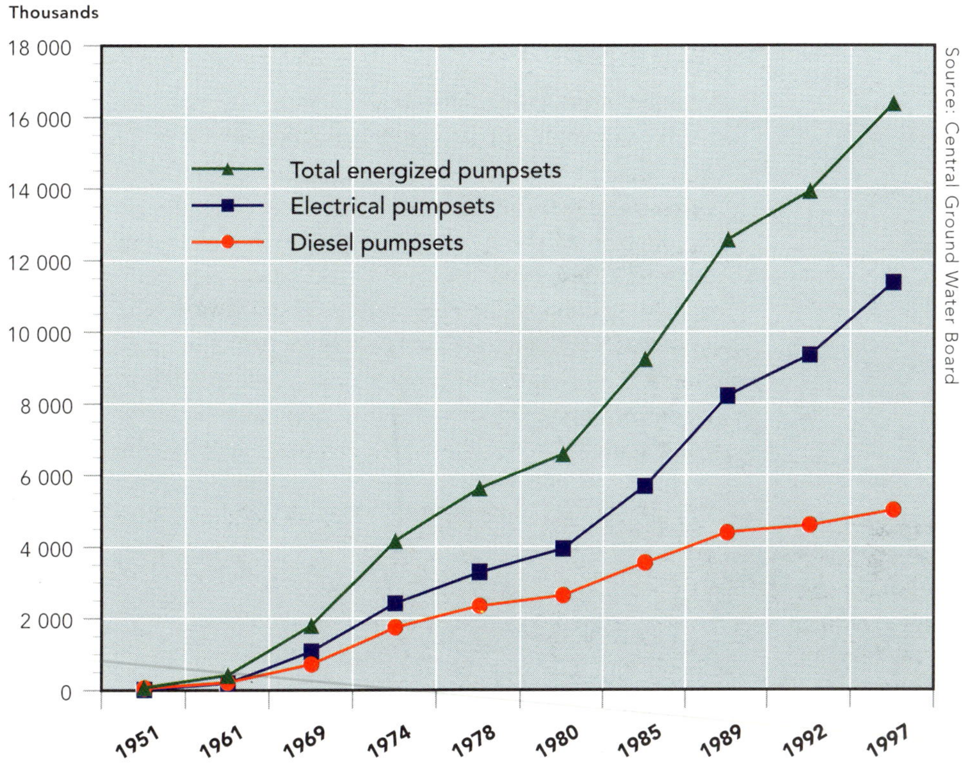

Direct human uses, however, comprise only a small fraction of the values dependent on groundwater. Base flows in streams, wetlands and surface vegetation are in many cases dependent on groundwater heads (or levels) and discharges. Change in those levels or changes in groundwater quality induce ripple effects through terrestrial and aquatic ecosystems. In some cases, marine ecosystems, such as those associated with submarine springs, may be supported by groundwater outflows from adjacent land masses.

In many countries, groundwater extraction has increased exponentially with the spread of energized pumping technologies for irrigation in agriculture. In India, for example, the number of diesel and electrical pumps jumped from 87,000 in 1950 to 12.6 million in 1990 (Central Ground Water Board, 1996, p. 9) and is illustrated in figure 1. Expansion of pumping technology has often resulted in dramatic declines in the water-table in low-recharge areas. Competition among agriculture, domestic and commercial water users over access to limited groundwater resources is growing. Perhaps more seriously, pumping changes flow patterns, often resulting in migration of pollutants and low-quality water into aquifers. Pollution has also been caused by rapid growth in the use of agricultural chemi-cals and by the all-too-common practice of discharging untreated industrial and domestic waste water directly into the ground. Furthermore, even minor spills of some industrial chemicals, such as organic solvents, can cause large-scale groundwater pollution. Once aquifers are polluted, their clean-up can be technically impossible or simply uneconomic.

Pollution and declining water levels represent direct threats to the sustainability of environmental, domestic, agricultural and industrial uses dependent on groundwater. In addition, as demands grow and the limits of sustainable extraction become evident, competition between agricultural and other users is increasing rapidly. Competition over shared aquifers between countries is also becoming apparent. This can generate a "race to the pumphouse". Users extract as much groundwater as possible in order to maintain their rights and capture benefits for themselves before the resource is exhausted. The net result can be a spiral of growing demands and decreasing availability. It is important to recognize, however, that *overdraught and water-level declines typically affect the sustainability of uses that are dependent on groundwater long before the resource base itself is threatened with physical exhaustion.* Many uses and environmental values *depend on the depth to water*—not the volumetric amount theoretically available. Depth to water affects the economics of groundwater extraction, and declines can exclude the poor from access long before they exclude the rich. Water-level declines can also cause wetlands and stream flows to dry up even when the total amount of groundwater stored in a given basin remains huge. *In many ways, the sharpest points of competition between uses may have to do with the objectives of management, not with allocation of the volumes of water available.* This consideration leads to an even stronger definition of groundwater as a form of "critical capital" (Dubourg, 1997) and further heightens groundwater's instrumentality with respect to surface water.

The proximate causes of groundwater depletion and pollution, indicated above, are rooted in population growth, economic expansion, the distorting impacts of subsidies and financial incentives, and the spread of energized pumping technologies. These immediate causes mask deeper forces. In most regions, groundwater is not an inherently scarce resource. Emerging problems relate more to management issues—use efficiency, allocation and understanding—than to the ultimate sustainability or carrying capacity of the resource base. This in no way reduces the severity of emerging problems or the impacts they can have on society and the environment. It does, however, point towards the complications society faces in organizing management responses. To be effective, approaches have to respond to the deeper factors underlying emerging problems, not just to the proximate causes.

1.2 The nature and value of groundwater

Groundwater is, in most situations, a *common property resource* with extremely high use value.[4] It is also inherently vulnerable. Disposal of waste to aquifers and groundwater extraction affect neighbouring users in ways that are often difficult to predict and quantify. Conditions vary greatly between locations, and data, information and understanding are often lacking or in a form that cannot be readily understood by non-specialists—who make up the bulk of the immediate users. As a result, there is little public awareness of groundwater, the benefits it confers and the limits under which it is available.

[4] The term "common property" refers to the status of groundwater as a resource to which all overlying landowners generally have access. The term is not intended to reflect the legal status of groundwater (which varies greatly) as a "public", "common" or "private" resource.

Understanding the physical complexity of groundwater resources is fundamental in developing management approaches. In many parts of the world, people view groundwater as existing in an infinite "lake" or "bowl" underground. This misconception lies at the root of much public misunderstanding regarding the nature and implications of groundwater development. Groundwater occurs in aquifers, or, broadly, geological formations capable of producing usable amounts of water (Lloyd, 1981, p. 7). Aquifers are rarely homogeneous and their geological variability conditions the nature of groundwater flowing through their respective lithologies and structures. The greatest variations in groundwater flow patterns occur where changes in rock types—for example, limestone overlying sediments and a hard crystalline rock—induce discontinuities in flow and may bring groundwater flow to the surface on the junction between the two rock types. Here the authors' preference is to define the bounding geological formations that attenuate or restrict groundwater flow in aquifers as "aquitards" (as opposed to "aquiclude" or "aquifuge"), since it is rarely possible to determine zero-permeability of any geological formation. The arguments surrounding the deep disposal of nuclear waste alone are evidence of this uncertainty (Hazeldine and Smythe, 1997).

Within an unconsolidated alluvial aquifer, for instance, great lateral variations occur in the mix of gravel, sand and clay making up the aquifer matrix. These changes cause continuous variation in groundwater flow and aquifer responses to pumping but may be repeated sufficiently at small scale to allow the aquifer to be considered and pumped as a homogeneous unit. In larger-scale alluvial aquifers, layers of sand or gravel-rich sediment interbedded with clay-rich layers induce lateral flow following the more permeable sand-and-gravel-rich zones. At the scale of large sedimentary basins, vertical flow is reduced by clay-rich or other low-permeability layers (aquitards) and underlying groundwater flows under pressurized (or semi-confined) conditions (see figure 2 on the following page).

The interaction of aquifer material and groundwater flow is never straightforward and can involve several orders of differentiation. Even when an aquifer is considered homogeneous in lithological terms, the hydraulic parameters conditioning the flow of groundwater in all three dimensions can vary—anisotropy.

Mpongwe Block, Zambia.

The flows discharging from a block of dolomite at the junction with the surrounding shales is clearly visible in this Landsat 3 image.

Source: Geosurvey International

Figure 2 Aquifer heterogeneity and anisotropy

The distinction between heterogeneity and anisotropy is illustrated by this example from Alberta, Canada.

The range of values for K (m/day) in all three dimensions can be seen in the inserted table.

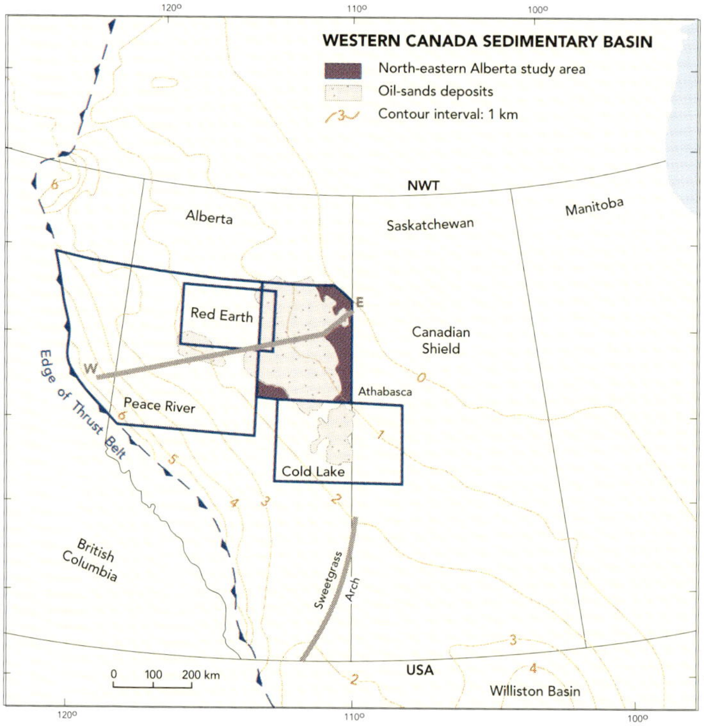

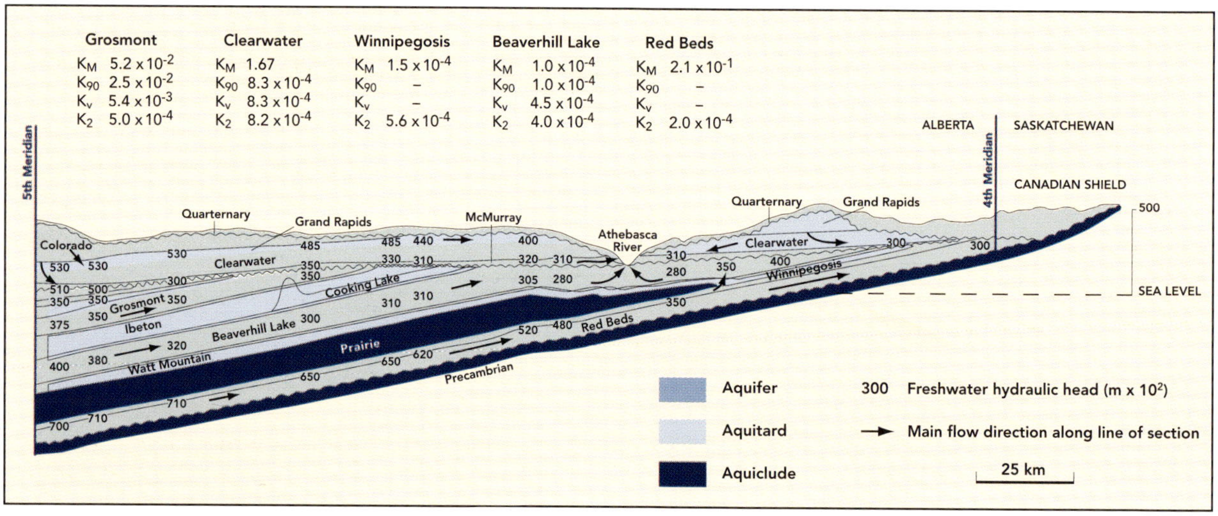

Hydraulic parameters for hydrostratigraphic units in the Cold Lake area, Alberta, determined from core analysis

K_M: maximum horizontal hydraulic conductivity
K_{90}: horizontal hydraulic conductivity at 90° angle with K_M
K_v: vertical hydraulic conductivity
S_s: specific storage
Median: 50 per cent value on the log-normal frequency plot for K and on the normal frequency plot for S_s
Hydraulic conductivity values are given in (m/d), and specific storage values given in (m^{-1}).

Source: Alberta Research Council

Figure 3 Migration of pollutants and low-quality groundwater under conditions of rapid urban growth

The arrangement of aquifers and aquitards can influence the impact of pollution dramatically and spread the hazard far away from the polluter.

Migration of pollutants

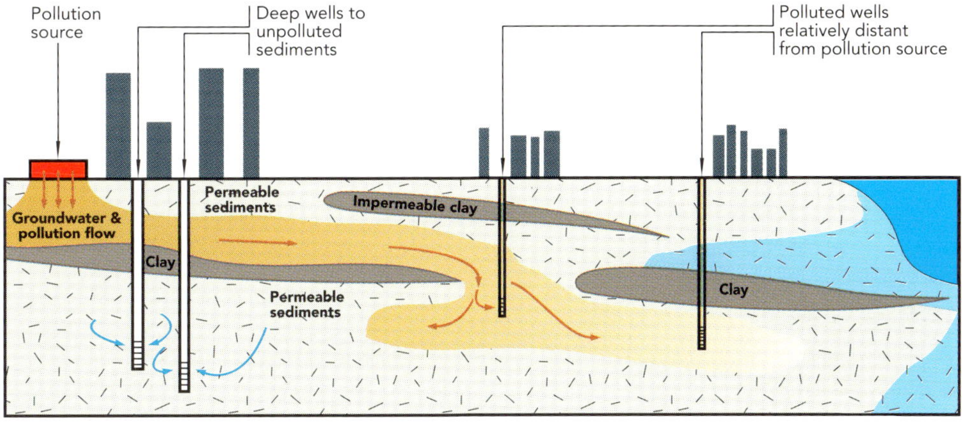

Source: British Geological Survey

Anistropic flow conditions are common in crystalline rock, where major fault zones often serve as conduits for or obstacles to groundwater flow. At a more micro level, rocks are often systematically fractured, enabling groundwater to flow easily in some directions and not in others. The anisotropic characteristic of both sedimentary and crystalline aquifers further complicates understanding and resultant management. Hence, heterogeneity and anisotropy greatly complicate groundwater management, since problems, as well as solutions, will be strongly influenced by specific local conditions.

The complex nature of groundwater is compounded in the context of pollution and quality problems (see figure 3). The chemical characteristics of aquifer materials and the way pollutants react with them vary greatly. In some cases, pollutants are "filtered" out mechanically or through adsorption onto particles within the soil or aquifer matrix. In other cases, however, pollutants remain mobile and can rapidly spread throughout an aquifer. The aquifer matrix itself can become contaminated and pockets of pollutants can serve as continuous sources of contamination. For example, small pockets of organic solvents can remain as pollution sources virtually indefinitely because of their low solubility in water. Furthermore, changes in pH or other groundwater characteristics can cause the release of toxic materials—such as fluoride—from natural sources within aquifers. Given the hundreds of thousands of naturally occurring compounds in groundwater and aquifer materials, and the similarly large number of compounds present in waste water released to aquifers, understanding and managing pollution problems is a highly complex task. This complexity illustrates the importance of avoiding pollution of groundwater from the start, rather than assuming problems can be addressed if needed after the fact.

Lack of awareness and understanding of aquifer systems, combined with the *common property* nature of the resource, perpetuates chronic *undervaluation of the resource base.* Because few understand the complex nature of groundwater flow and contamination, the vulnerability of the resource base to irreversible damage through unconstrained use or waste disposal is rarely appreciated. At the same time, because the groundwater resource base common to large areas can easily be tapped through wells on small individual landholdings, its condition is dependent on the actions of many users. Unless each individual can be assured that others are behaving in ways that protect the resource base, the *in situ* value of the resource to individuals is low. Finally, because the full value of the resource is rarely reflected in the costs individuals face when using it, they are rarely aware of that value. As a result, individuals are often indifferent to the source of the water they are using.

Very few formal attempts have been made to estimate both the immediate use and *in situ* values of groundwater. Economic or legal mechanisms through which *in situ* values could be communicated to groundwater users have yet to be systematically realized. Indeed, the first major publication on the issue has only just been published (National Research Council, 1997). One of the reasons valuation issues are complicated is that many of the services provided by aquifer systems (such as base flow in rivers) depend on *in situ* system characteristics and not just on the volume in storage or uses involving groundwater extraction. While there is scope for the direct-use values to be documented through economic studies or communicated to users through the operation of markets, documentation of the *in situ* values associated with groundwater systems is much more complicated, and these values are rarely reflected in market transactions.

The development of formal water markets has been a major focus of experimentation and analysis in countries such as the United States. In many other countries, however, informal markets and trading in groundwater have developed apace with the emergence of mechanized wells (Shah, 1993). While these markets often at least partially reflect use values, market development has generally occurred in the absence of any regulatory or physical planning framework (Solanes, 1996). Even where emerging problems are recognized, governments and other organizations often lack the capacity to control use or extraction directly. These underlying factors have allowed extraction and pollution to proceed unchecked as demands grow and pumping technologies spread.

1.3 Integration of physical and socio-economic systems

The above context suggests that emerging groundwater problems need to be addressed through integrated management approaches designed to change the way people view and use the resource. This involves an appreciation of three effective levels of *integration,* integration within the hydrologic cycle (the physical processes), integration across river basins and aquifers (spatial integration) and integration across the overall social and economic fabric at national and regional levels. Aquifer systems and subsystems are intimately connected with other portions of the overall hydrologic system and are best understood within broad physical frameworks upon which hydrologic and hydrogeologic processes are set (Alley, 1993; Jones, 1985; Burke, 1994; Burke, 1996) (see box 2). Similarly, the patterns of groundwater uses are interconnected and often sequential. Taken together, hydrology and use patterns form interdependent systems linked by the way water flows through the resource base and through each use. This meshing of physical and socio-economic frameworks is key to understanding patterns of resource use and management options. It is also often a complicating factor because the lack of congruence between natural boundaries and social or administrative/political boundaries can impede attempts to manage on the basis of natural units. There are very few instances where this type of integration has taken place. The classic "Water Resource Master Plans" developed in Africa, for example, during the 1980s give some indication of the type of spatial "overview" of physical and socio-economic systems that have been used to develop water-supply and sanitation programmes (e.g., Norconsult, 1982; United Nations, 1990). More recent attempts to use geographic information system technology to overcome scale and integration problems at the national level are gaining ground (United Nations, 1998b). However, the degree to which these initiatives, dependent primarily on external support, have been internalized and used to drive sustainable approaches to groundwater development has varied. More recent approaches to groundwater resource development and management have sought closer integration with local socio-economic groupings. The "gestion des terroirs" approach in Burkina Faso and the use of public "water forums" in groundwater and surface-water planning and development at sub-catchment and basin scales are cases in point (United Nations, 1995a).

The interdependent nature of hydrologic, hydrogeologic and water-use systems makes the continuum between data, information and knowledge of particular importance. Understanding systemic interactions is essential as a basis for identifying management options and generating sufficient social consensus to implement them. Developing this understanding requires a steady flow of hydrologic data and data on water use. It also requires continuous refinement of the scientific foundations, both physical and social, upon which management solutions rest. There are few examples of this type of progressive approach to groundwater information and societal needs. The Groundwater Forum in the United Kingdom (Grey et al., 1995) and the Groundwater Foundation in the United States represent initiatives designed to address, respectively, groundwater research needs and public education in developed countries. The integrated resource planning process recently undertaken by the Metropolitan Water District of Southern California represents one of the few large-scale examples where organizations have attempted to implement this type of approach in specific planning contexts (Metropolitan Water District, 1995). But the more pressing needs for this approach are in developing countries where sophisticated professional associations, well-organized user groups, appropriate government agencies and information flows are often absent. Problems are particularly severe where well-organized special interests promote self-serving policies in the absence of the transparent governmental and information systems essential for other stakeholders to counterbalance their influence. In such cases, as in the example of Mendoza in Argentina referred to above, participation by a narrow selection of stakeholders becomes a vehicle for "pork-barrel politics", at the expense of both the resource base and social equity. The negative environmental and social impacts of rent-seeking stakeholders benefiting from privileged access to the decision-making process are poorly documented but probably pervasive. Even widespread public participation is no panacea. Hydrologic complexity and poor public understanding of groundwater systems have, in some cases, led to widespread public support for—and the implementation of—management approaches that are technically questionable (Wilson et al., 1998).

Box 2 Development of physiographic frameworks for understanding groundwater systems: the Kafue sub-basin, Zambia

The Kafue sub-basin has an area of 155,000 km^2 and occupies 20 per cent of Zambia's total land area. It plays a central role in Zambia's economy. It forms one of the principal subcatchments of the Zambezi basin and has a mean annual flow of some 300 cubic metres per second near its confluence with the Zambezi. The mean annual flow represents only 6.2 per cent of the mean annual rainfall of 1,057 mm falling over the catchment area. This low-yielding hydrological regime of the Kafue is greatly influenced by the geological and geomorphological setting of the basin. The topography is very subdued, with the landforms being developed largely by chemical weathering and erosion. This suggests a very active and widespread role for groundwater circulation in the hydrology of the Kafue basin. High open-water losses from areas of impeded drainage and high evapotranspiration losses from groundwater seepage zones are associated with saprolite and karstic aquifers. Despite the detailed hydrogeological work in the immediate post-independence years, the late 1960s and early 1970s, plans for the basin's development are based on a very limited understanding of the basin's combined surface-water and groundwater resources. Water-supply schemes have generally concentrated on stored surface water or run of river solutions and ignored the potential of some very prolific and ideally located aquifers. This has resulted in suboptimal use of the system's resources and disputes between user agencies during dry-season recession. Equally, measures to protect the open dolomitic aquifers from urban, industrial and mining pollution sources have been progressively bypassed. A brief examination of the physical background of the Kafue catchment reveals a geological/geomorphological framework in which to discuss water resource development and management. The figures illustrate a set of principal subcatchments with well-defined geological, geomorphological and hydraulic characteristics which, contributing to the catchment hydrological response, provide a template in which to discuss water resource planning. Without such a framework it is difficult to explain adequately the hydrological response of the catchment, to account for the distribution of the available surface-water and groundwater resources and to plan development and appropriate protection measures.

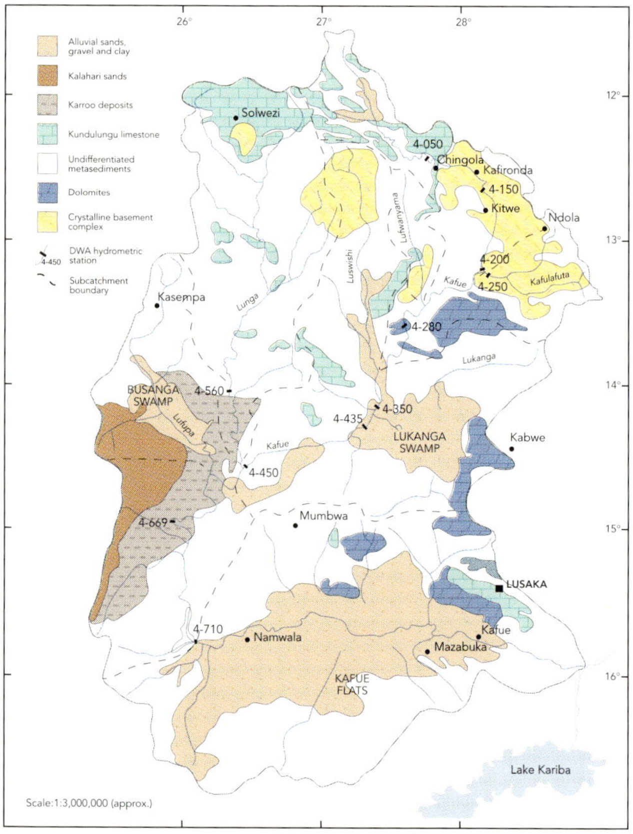

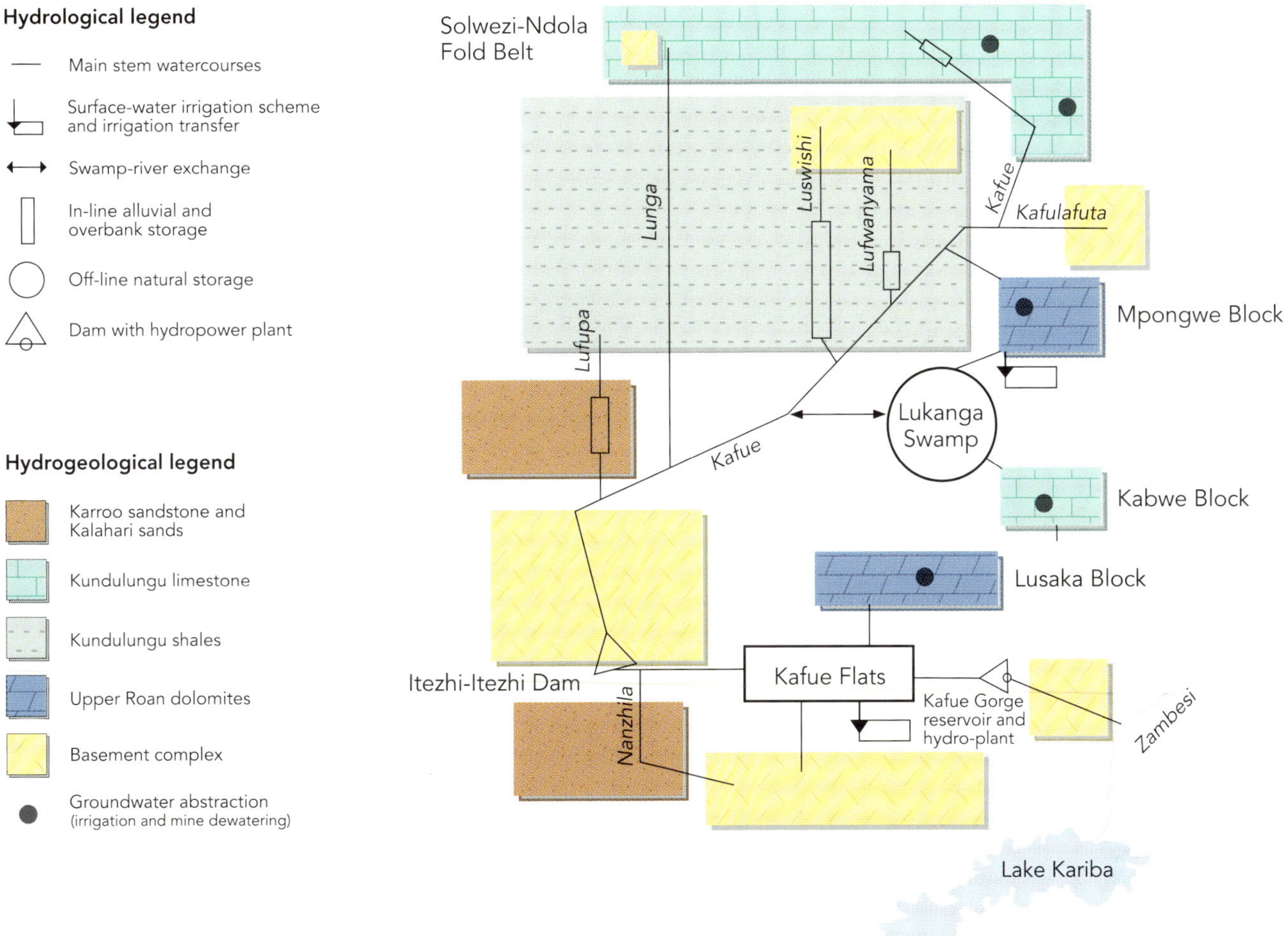

Source: Burke (1994)

Interdependency also makes integrated approaches to management essential. This, in turn, depends on the availability of political and socio-economic frameworks that are capable of responding sensibly to the resource's physical structure and dynamics. Such frameworks consist of the legal, institutional and regulatory structures that enable or obstruct different forms of organization or types of action. Community participation in developing groundwater management solutions needs to be widespread, given the highly distributed nature of groundwater resources. Physical management options vary greatly at a local scale. Furthermore, the large number and dispersed nature of groundwater users can limit regulatory avenues. Legal and institutional frameworks need to reflect this by enabling different types of community action in response to differences in physical management options and needs.

The importance of participation does not reduce the role of regulatory or legal rights systems that govern groundwater access and use. Rights, regulation, participation and education all represent different dimensions of a single equation. Individuals are often unwilling to participate unless regulatory or legal structures are available to control those who may be less willing to cooperate. At the same time, regulation is often resisted unless the need for it is understood and accepted and unless individuals are assured that regulatory approaches reflect fundamental rights. Frameworks, as a result, need to contain mechanisms that both enable regulation or legal controls over groundwater and explain the necessity for those controls. Culture and local traditions are also important factors. In some areas, these provide a sufficient basis for development of decentralized participatory institutions; in others, more centralized systems are viable.

Finally, it is essential to recognize the influence of economic factors. Formal and informal markets are often major mechanisms for water allocation. Prices established in these markets can encourage or discourage use efficiency. Apparently indirect policies, such as those affecting energy prices and the prices of agricultural products, can also significantly affect groundwater-use incentives.

In conclusion, frameworks and processes that enable integrated packages of participatory, regulatory, economic and other interventions to be developed and implemented are essential. These frameworks and structured interventions are fundamental to the sustainable development of all regions dependent, even partly, upon aquifer systems.

2 The role of groundwater in society

Why should we care about groundwater? The factors cited in the introduction capture some imperatives at the global scale. The core motivation, however, lies closer to the ground and relates to the dynamic nature of development and the aspirations towards environmental sustainability. "Global pictures" are difficult to relate to the wide variation in groundwater conditions and problems encountered by people in their everyday lives. Groundwater resource conditions directly affect the opportunities and risks people face and the environment they live in. Furthermore, the keys to groundwater management generally lie in the local realities of hydrogeology and the sociocultural patterns of water use, not in global statistics, agreements or statements. The details of how people organize themselves around groundwater are particularly important, more so than for surface water, because of the direct and intimate physical links that are developed and because of the particular jurisdictional and institutional responses this use has invoked (Barraqué, 1997b).

Understanding the role groundwater plays as a foundation for the daily lives of people around the world requires an understanding of the dynamics of real situations in local contexts. The sections below outline some of the most important dynamics and locate them within the larger picture provided by higher-level concerns.

2.1 General socio-economic development

Groundwater is a key input to agriculture, health, industry and the environment. In many ways the role groundwater plays in socio-economic development is intangible—yet it lies at the root of most human economic activity. People have always chosen to settle in locations with reliable water supplies. Without the initial decision—heavily influenced by access to water—that a location is appropriate for settlement, socio-economic development would not have begun. Once the initial decision is taken, water plays a critical role at all stages in the growth of socio-economic capital. From drinking supplies or the mixing of mortar for buildings to irrigation, water is an essential input. Reliable water supplies, which are generally either directly or indirectly derived from groundwater, carry a special premium, since water-supply reliability is a key parameter influencing the sustainability of essential activities. As the reliability of water supplies declines, risks for users dependent on them increase. From small investment decisions by poor farmers regarding the purchase of fertilizer or seeds to major investments by companies or govern-

Agriculture in Gujurat is dependent upon access to groundwater.

ments, economic returns (and in some cases survival) depend on whether or not water is available when needed. In many areas, the assurance provided by groundwater is, as a result, a key foundation for socio-economic development.

While it is essential to recognize the role groundwater plays as a basis for socio-economic development, it is equally important to recognize that all uses produce inherent externalities, or have negative impacts upon the resource base. Most uses of groundwater are consumptive or involve a degradation of the quality of the water that is returned to the aquifers by recharge. The stock of groundwater is generally reduced and the subsequent quality in the all-important shallow aquifers is degraded. In the longer term, prolonged abstraction beyond the rates of recharge will involve fundamental changes to the configuration of aquifer systems. This interdependency of uses and impacts is a key factor that should be kept in mind throughout subsequent discussions of the key services groundwater produces and the problems facing sustainable management of these resource bases.

2.2 Key groundwater services

2.2.1 Food security

Over the past 50 years, expansion of groundwater irrigation has played a lead role in food security. Yields in areas irrigated by groundwater are often substantially higher than yields in areas irrigated from surface sources. In India, for example, research indicates that yields in groundwater-irrigated areas are higher by one third to one half than in areas irrigated from surface sources (Dhawan, 1995). As much as 70-80 per cent of India's agricultural output may be groundwater-dependent (Dains and Pawar, 1987).

Higher yields from groundwater-irrigated areas are due, in large part, to its ease of control and reliability. Early studies indicated that water control alone can reduce the gap between potential and actual yields by about 20 per cent (Herdt and Wickham, 1978). In many cases, groundwater has a specific advantage over surface water for irrigation—it is generally on demand and "just in time". Taken with the esentially private and decentralized nature of the resource, this can permit highly efficient and flexible in-field application, which can translate into substantial benefits for groundwater users. Barraqué, for example, notes that "irrigation uses 80% of all water in Spain and 20% of that water comes from underground. The 20%, however, produces more than 40% of the cumulated economic value of Spanish crops" (Barraqué, 1997b). Reliability is even more important. Groundwater is a key buffer against drought and normal variations in rainfall. Yacov Tsur, an agricultural economist, estimates that the stabilization value of groundwater in agriculture is as much as 50 per cent of the total value of the resource in California and "more than twice the benefit due to the increase in water supply" in the Negev desert (Tsur, 1990; Tsur, 1993).

In general, increased yields from groundwater-irrigated areas have translated into substantially higher yields and are thus a major factor in food production at the regional and national levels. The significance of this should not be underestimated, since internationally traded grain is a small fraction of global consumption. The magnitude of grain transfers that would be required to meet major deficits in areas such as South Asia could exceed the capacity of international trading systems. In the face of a localized famine after the 1967-1968 droughts, the United States shipped 20 per cent of its wheat crop to India (Brown, 1970), but over the last decade, food grain imports to the region have been far less important. Much of India's ability to meet the food needs of its growing population is due to groundwater irrigation. Some of the most important food security benefits related to groundwater lie, however, at the level of individual farmers. The vulnerability to natural hazards of different groups in society, including those that threaten food security, can be explained by their access to networks of key productive and social resources (Blaikie and Canon, 1994; Ribot et al., 1996). For rural populations, groundwater is among the most important of these resources.

Households with access to key resources are able to build support systems that reduce their vulnerability to natural hazards. Groundwater irrigation reduces the risk that investment in labour, seed, fertilizers, pesticides and other inputs will be lost due to drought or the variability of precipitation in normal years. At the same time, higher yields from groundwater-irrigated areas enable households to generate surpluses. As a result, households with access to groundwater tend to have higher levels of savings and are able to make investments in other productive resources or activities (but the reverse is also true—households with better access to social and economic resources are generally better able to afford the cost of obtaining groundwater access, since private wells and boreholes can be expensive). When drought strikes or there is a gap in rainfall, these households have a dual advantage. First, having control over an assured source of irrigation, they are far less likely to suffer losses than those without access to groundwater. They may even be well positioned to improve their income through the sale of products at prices above those prevailing in non-drought years. Second, even if "the well runs dry", households that own wells have often been able to save cash or food and

Flowing artesian wells in karstic areas of Hunan Province, China, provide multiple household services.

invest in alternative sources of income. As a result, they have assets that can carry them through periods of scarcity or crisis. This second benefit gives households owning wells a comparative advantage in the face of many hazards—flood, disease and economic downturn—not just drought. In this sense, groundwater offers a means to manage environmental and economic risk. Access to groundwater is thus a key "enabling" factor increasing food security at the household level and reducing overall vulnerability. But here, again, the inherent nature of aquifer systems can complicate matters and will necessarily determine the particular socio-economic opportunity groundwater presents (see box 3).

Box 3 Basement complex aquifers in Africa: geomorphology and water resources

The inherent nature of aquifer systems can complicate matters. The thin, discontinuous aquifers of the crystalline basement complex in many arid and semi-arid settings in Africa and the Arabian peninsula are particularly susceptible to rapid drainage following recharge events and may not provide an inter-annual buffer. As rainfall increases on the basement complex and weathering products have a chance to accumulate as saprolite, the aquifer systems become more contiguous and the porosity of the saprolite clays offers a chance for inter-annual storage of recharge. This is a complex feedback process that occurs over space and time but has important implications for rural development, particularly in many parts of tropical and subtropical Africa (Jones, 1985; MacFarlane, 1985; Acworth, 1987, Wright and Burgess, 1992).

Groundwater occurrences on the older (higher-elevation) African erosion surfaces are linked to specific weathering processes and saprolite profile development. These processes are reliant upon sustaining groundwater circulation over crystalline basement complex lithologies through characteristic saprolite catenas. Weathering products have to be removed at rates sufficiently fast to develop and maintain the permeable granular basal weathering layer immediately above the fresh bedrock. This system will remain open as long as rates of recharge and discharge are maintained. If the rate of groundwater circulation is not maintained through reductions in recharge and/or through-flow, the weathering products are not removed, the system becomes closed and weathering at the base of the profile is severely attenuated. It is the movement of water in and out of the saprolite catenas that drives aquifer development, and to this extent the groundwater circulation acts independently of the range of lithological, hydrochemical and biological controls on the weathering processes.

Recharge zone

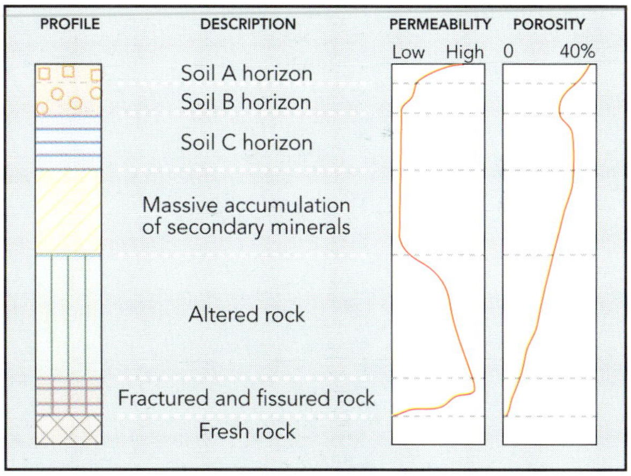

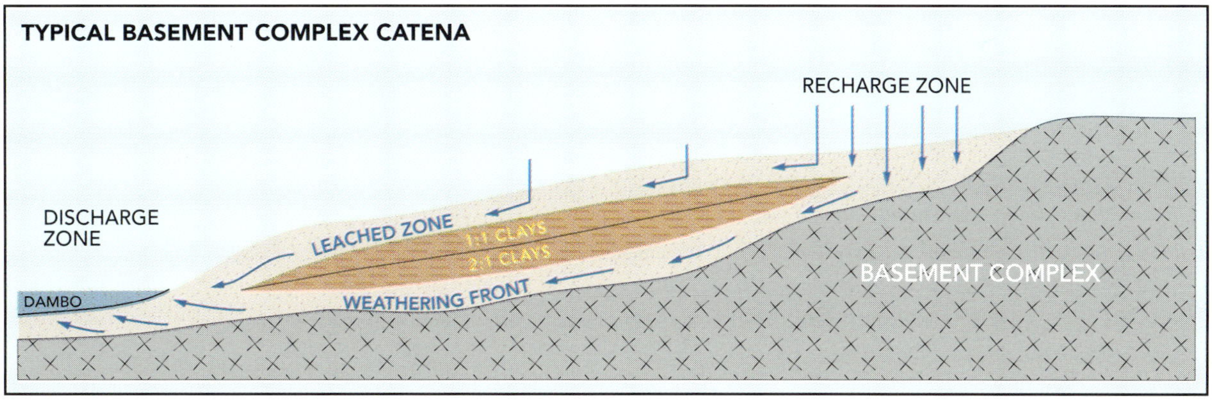

Both the macro and micro roles groundwater plays in decreasing vulnerability will be of particular importance in the context of global climate change. Most climate-change scenarios for semi-arid regions project increases in inter-annual variability, drought frequency and extreme events (Ribot et al., 1996, p. 14). The buffering role of groundwater in regional crop production and reduction of household vulnerability will tend to moderate the impact of increased fluctuations.

2.2.2 Domestic water supply

Groundwater is a highly important source of domestic water supply. In India, roughly 80 per cent of rural water supply for domestic uses is met from groundwater. In the United States, 36 per cent of the population receives drinking water from public systems using groundwater and 16 per cent from private wells. Reliance on groundwater for domestic supplies does not imply that domestic needs represent a large fraction of groundwater extraction. Although over half of the United States population meets all or part of its drinking needs from groundwater, this accounts for only slightly over 4 per cent of total groundwater use in

Guiyang, the rapidly expanding capital city of Guizhou Province, China, sits on the divide between the Pearl and the Yangtse river basins on vulnerable karst aquifers.

the United States (National Groundwater Association, 1996). In England and Wales, groundwater accounts for about one third of water extracted for public supplies (Grey et al., 1995, p. 4). In many arid countries, dependence on groundwater is much higher and it forms the primary source of urban and rural supply. For example, in Namibia, a country without a national hydrogeological map, up to 60 per cent of bulk water supplies is derived from groundwater, although the majority of water-supply investment is tied up in surface-water storage schemes, and in India, 80 per cent of rural drinking needs are supplied from groundwater.

Figure 4 Projected increases in urban population, 1996-2030

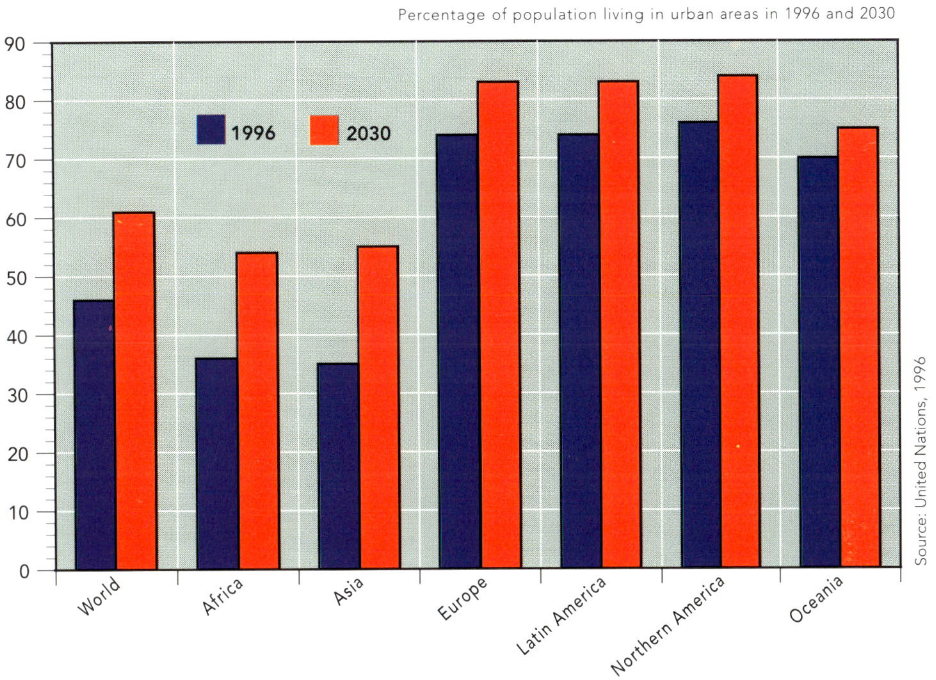

Dependence on groundwater for domestic supply is increasing. Urban populations are growing rapidly. In 1950, less than 30 per cent of the world's population lived in cities and towns, whereas now, at the end of this century, it is anticipated that this figure will have risen to 50 per cent (see figure 4). Where available, groundwater is often chosen as the primary source for urban and rural drinking-water-supply systems. Its ubiquity and generally high quality obviate the need for expensive treatment, so that the cost of procuring bulk water is significantly reduced.

The importance of potable drinking water is clear. As in the case of other uses, however, broad statistics of the type cited above paint only a limited picture of the value of groundwater as a source of domestic supply. Perhaps a more telling picture is captured by the reactions of locals to outside visitors. A glass of water is the first and essential symbol of hospitality offered to outsiders in many parts of the arid and semi-arid world. In India, any slight hesitation on the part of the recipient will often lead to a quick response: "*Yeh kuon ka panni hey, taanda hey* . . . (This is well water, it's cool) . . .". Well water is cool and, by implication, safe for anyone to drink.

The ubiquitous nature of groundwater has a major impact on the daily lives of people in many parts of the world. Wells in villages and towns free people, particularly women, from long daily walks to fetch water from springs or rivers for livestock and domestic uses. This frees time and labour for other activities. Furthermore, since water no longer has to be carried over long distances, more is often used. This can have major health benefits. There is a growing consensus that the amount of water available for personal and domestic hygiene is as important a determinant of a community's health as the quality of that water (Esrey and Habicht, 1986). As they indicate, "The provision and use of sufficient water, albeit of poor quality, could prevent the contamination of food, utensils and hands and thereby reduce the transmission of major infectious agents of diarrhoea" (Esrey and Habicht, 1986, p. 118). The minimum requirement for this is estimated at 20 litres per capita a day. Above 20 litres per capita a day the health benefits of additional supply decline and the importance of water quality increases. Water-borne pathogens and pollutants are among the major causes of disease. Diarrhoea, for example, caused an estimated 4.6 million deaths a year during the late 1980s in developing countries excluding China (Gleick, 1993, p. 205). Groundwater is commonly much cleaner than surface-water sources because of the ability of soils and the aquifer matrix to filter and diffuse pathogens and pollutants. Access to groundwater often enables populations to increase both the volume and quality of the water they are able to use.

2.2.3 Equity, poverty alleviation and rural development

Groundwater is a key resource for poverty alleviation and the economic development of rural areas. The ubiquity and, in areas with high water levels, the relatively low cost of obtaining access to groundwater are also major factors contributing to poverty alleviation through access to irrigation and improvements in health and household sanitation (see box 4). Evidence indicates that improved

The relative wealth of communities in the Hadramaut valley, Yemen, is determined by the productivity of a superimposed aquifer system.

water sources generate many positive externalities in the overall household micro-economy. In Africa, evidence indicates that potable supply, sanitation, stockwatering and garden irrigation are all equally important, particularly in agro-pastoral communities in arid regions where communities may rely totally on groundwater (see box 5). It should be noted that there is a strong spatial correlation between population density and groundwater use in both developed and developing areas of the world. This points to the continued reliance humankind will continue to have on groundwater for overall development, but also indicates the concentration of the negative externalities associated with groundwater use—drawdown, pollution, mineralization and subsidence (Shah et al., 2000).

Life in periurban communities in Malawi is characterized by the daily task of collecting water from shallow, polluted aquifers.

In areas dependent on irrigated agriculture, the reliability of groundwater sources and the high crop yields generally achieved as a result often enable farmers with small landholdings to increase income. In India, small and marginal farmers (those having less than 2 hectares) own 29 per cent of the agricultural area. Their share in net area irrigated by wells is, however, 38.1 per cent, and they account for 35.3 per cent of the tubewells fitted with electric pump sets (Government of India, 1992). Thus, in relation to operational area, small and marginal farmers tend to have proportionally more irrigated land than larger farmers. With productivity on irrigated lands being much higher than that on non-irrigated tracts, better access to irrigation for small and marginal farmers can significantly reduce poverty in rural areas.

Box 4 Households, risk and poverty

Access to groundwater for a poor rural family represents a point of certainty in a world otherwise fraught with risk. Without the security of a well, investments in seed, fertilizer and pesticides carry a high risk. In purely rain-fed areas, such investments can easily be lost if the rains do not come on time. Even where irrigation from surface systems is available, individual farmers often have little control over the amount and timing of water deliveries. As control over water decreases, the risks associated with each additional investment to increase crop yields grow. Losses can be devastating. Small farmers throughout the developing world often depend on high-interest loans from moneylenders for critical agricultural inputs. Crop failure often throws farm families into permanent debt and can result in the loss of land and force them into positions as indentured agricultural labourers.

Access to groundwater greatly reduces the above risks. In addition, it can enable farmers to slowly lift themselves out of poverty. Control over water, the lead input to agricultural production, enables farmers to fine-tune irrigation to crop needs. Yields are, as a result, higher. This itself often enables farmers to increase income and living standards. More importantly, however, the surpluses generated greatly reduce the vulnerability of marginal farmers (Blaikie and Canon, 1994). Surpluses enable diversification. Children can be educated and non-agricultural-income activities initiated. Accumulated resources are available to meet needs in the face of economic fluctuations and natural or other disasters. In sum, the overall position of small farmers to face the wide variety of challenges encountered in their lives is greatly enhanced by access to groundwater.

As the photo illustrates, even low-quality groundwater lenses in the atolls of Kiribati can at least provide water for washing.

The positive economic impact of groundwater development extends beyond well owners. Studies in Pakistan have documented the importance of groundwater irrigation to those who do not own wells (Meinzen-Dick, 1996). In these studies, farmers owning wells generally achieved the highest yields, while those purchasing water from well owners achieved yields higher than those dependent on canal irrigation alone, but not as high as the yields achieved by well owners. In addition, those purchasing water tended to have higher fertilizer, labour and other inputs than those dependent on canal water alone. This stabilizes the demand for these associated inputs, and leads to the spread of support services for pumps and wells, creating a base for small-scale rural industries. Furthermore, the spread of groundwater irrigation can increase demand for labour. In India, for example, labour accounts for approximately 44 per cent of the cost of installing a well, and the additional indirect employment created on every hectare of irrigated land through increased agricultural activity is approximately 45 days per hectare (Government of India, 1996). Therefore, the expansion of groundwater irrigation has significant ripple effects, creating employment throughout rural economies.

The equity impacts of groundwater development for irrigation are, however, not all positive. Modern tubewell and drilling technology tends to be capital-intensive. As a result, early exploiters of groundwater have typically been large farmers who produced surpluses for the market. This has been the case in Western Europe and the United States, and in parts of Africa where large-scale commercial farming based on groundwater, such as on the dolomitic aquifers in Zambia, has been predicated on the complete or "perfect" ownership of groundwater rights linked to the large landholdings. Equity implications of the link between ground-

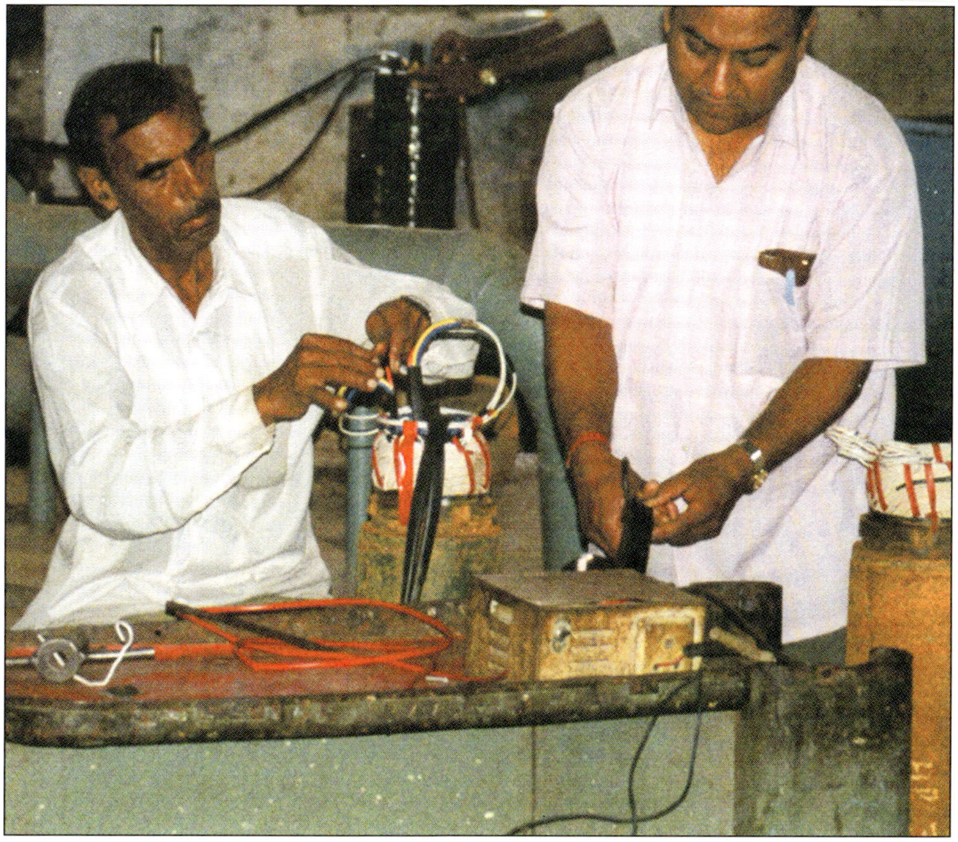

Pump service technicians are an important element of rural communities in India.

Box 5 Access to groundwater in central Sudan

The development of groundwater in the undulating terrain of central Sudan has been traditionally reliant upon shallow groundwater circulation in the superficial alluvial deposits and the accumulation of surface run-off in hafirs. Satellite imagery reveals a complex mosaic of recharge styles and groundwater-supported seeps in a range of sedimentary and crystalline aquifers.

Mechanized boreholes now tap the moderately transmissive Umm Ruwaba and Nubian sandstones in a set of faulted sedimentary basins and the fractured and weathered crystalline aquifers on the margins of structural culminations (Burke, 1994). This has permitted a rapid growth in rural populations across the semi-arid belt in Dafur and Kordofan. It has also encouraged rapid growth in the traffic of stock from the grazing lands in the south to the markets in the north. Stock routes with wateryards and veterinary services were established in the 1980's to support this growth in traffic (World Bank, 1984). The cattle that were once dependent upon the seasonal occurrence of shallow groundwater are now able to survive the long trek to distant markets.

Access to assured groundwater supplies is now critical for both pastoral and sedentary populations and has resulted in tension between established communities and the relatively wealthy pastoralists and stock traders moving stock to the markets around Khartoum. The communities that once followed the seasonal pattern of rainfall are now fixed and reliant upon recently constructed wateryards and have to compete economically with the pastoralists for access to the wateryards designed for stock water use—sometimes exploiting groundwater with very high levels of salinity. These economic opportunities and the issues of access and the attendant problems of land degradation around the wateryards have been brought on by technology.

Traditional water-lift structure in south Yemen exploiting wadi underflow

water and landownership are particularly important for countries with small, fragmented landholdings. Smallholders growing subsistence crops often depend on supplementary groundwater irrigation using a variety of man- and animal-driven water-lifting devices from shallow, open wells. The expansion of energized pumping technologies tends to draw water levels down, driving shallow wells and muscle-driven devices out of business. This was, for instance, the case throughout the Gangetic basin and other parts of India during the 1960s. As water levels began to decline, some state governments in India attempted to implement administrative regulations, such as selective credit controls, restrictions on electricity connections and siting and licensing rules. These regulations did not affect landowners who were able to install tubewells early, but limited the entry of latecomers—particularly the resource-poor who depended on credit or access to subsidized electricity in order to afford the capital cost of operating and installing pumps. In addition, the economically and politically powerful were generally able to bypass regulations via "adjustments" with officials or by depending on their own financial resources for well construction and operation.

This scale bias has attracted widespread policy attention in South Asian countries, where groundwater irrigation has grown rapidly in recent decades and now surpasses canal irrigation. Between 50 and 60 per cent of Indian irrigation—and much more in Bangladesh—depends upon groundwater. Many of the subsidies intended to encourage groundwater development by small farmers have, however, gone to the benefit of a rural elite (Mundle and Rao, 1991; Kahnert and Levine, 1989). The most marginal farmers still lack the capital necessary to afford access.

Equity considerations are generally a major point of tension as management needs emerge (see box 5). Rapid unrestricted development of groundwater has reduced poverty in many parts of the world by giving the poor access to a key resource for production. This same pattern of unrestricted development, however, is the primary cause of over-extraction and quality problems now emerging in many parts of the world. As groundwater problems grow, marginal populations are often the first affected. Water-level declines, for example, have the largest economic impact on individuals who are unable to afford deeper wells—i.e., the poor. The poor are also the least well positioned to protect their interests if groundwater extraction must be reduced. They are, for example, often the last to construct modern wells and boreholes. As a result, restrictions on new wells tend to affect them much more than wealthy communities, where wells were installed much earlier. Wealthy individuals and communities are also often able to work around these and other types of management restrictions, while poorer communities (who generally lack political as well as economic leverage) have less ability to do so. In sum, there is an inherent tension between equitable access to groundwater for all sections of society and sustainable management of the resource base.

2.2.4 Environment

The array of environmental services or values dependent on groundwater are often poorly understood. Environmental concerns related to groundwater generally focus on the impacts of pollution and quality degradation on human uses, particularly domestic supply. Development impacts on the groundwater environment are, however, different from the numerous environmental services provided by groundwater resources in their natural state (see box 6).

What are some of the most important environmental services provided by groundwater? The following list, while not exhaustive, illustrates the degree to which many environmental values are dependent on groundwater:

- ***Catchment base flow*** derived from groundwater discharge is perhaps the most evident environmental value associated with groundwater. In many areas, springs and the dry-season flow in rivers depend heavily on groundwater. This is important in both humid and arid regions where precipitation is highly variable. Between precipitation events, groundwater and return flows from agricultural, domestic and other users are the primary source of flows in rivers. Since return flows generally have higher pollution loads than groundwater flow, the groundwater contribution is important to both the quantity and the quality of dry-season flow in surface watercourses. Beyond this, groundwater levels are a controlling factor governing the extent of wetlands and surface vegetation types.

Base flows in Nepal maintain year-round water services.

Box 6 Protecting groundwater environments: the San Luis Valley, southern Colorado

The San Luis Valley (SLV) in southern Colorado is characterized by diverse agricultural, economic, environmental and cultural systems that are sustainable, compatible and interlinked by patterns of water use. Return flows from surface irrigation systems and streams are a primary source of groundwater recharge in the SLV. These flows help to maintain high water-tables in many parts of the valley and counterbalance the effect of extensive pumping for irrigation and, to a lesser extent, domestic water supply. In-stream flows and vegetation communities in many parts of the SLV depend on the high water-tables. In addition, the high water-tables sustain a distributed and interconnected network of seasonally flooded wetlands which are among the best in the inter-mountain West. Numerous natural heritage sites containing endemic and rare species have been identified by the Colorado Natural Heritage Program within the SLV. As one review concludes, "Protection of the unique hydrological regime of the Valley and the maintenance of the water-table will be necessary to conserve both riparian and terrestrial sites and habitats" (Pague and Simonson, 1994).

In addition to the environmental values, established communities depend on current patterns of water use. Cultural roots are deep in the SLV and derive from early Hispanic communities and later Mormon and other Anglo settlers. It is among the oldest permanently irrigated areas in Colorado. Irrigation started in the 1880s, and "by the year 1900, 1800 miles of canals, ditches and laterals had been constructed in the valley, and the streams flowing into the valley were essentially fully appropriated" (McFadden, 1989, p. 111). Now, irrigated agriculture "is the basis of a traditional way of life for ethnic communities that are proud of their histories and have a high level of interest in maintaining their historic way of life despite pressures for change" (National Research Council, 1992, p. 168). From economically sophisticated family-owned operations to small farmers and ranchers, all communities in the SLV depend on irrigated agriculture.

Proposals to export water from the San Luis Valley to meet the growing demands of the

Denver metropolitan area and Albuquerque have emerged with growing regularity over the last decade. If implemented, these proposals could disrupt the delicately balanced hydrologic equation that lies at the heart of environmental, economic and cultural systems in the SLV. Most debates over water transfers from the SLV focus on potential direct impacts on specific high-priority environmental sites or the water rights of specific other users. Potential systemic impacts are rarely recognized. In addition, even where the interlinkages are recognized, no clear legal mechanisms exist to incorporate them in the formal decision-making process over transfers. This issue is likely to emerge in many situations where water management decisions have the potential to affect widely distributed sets of environmental, cultural and economic values.

- ***In-stream fisheries and aquatic ecosystems.*** In-stream flows are critical for the maintenance of fisheries and aquatic ecosystems. As previously noted, groundwater contributions can be a dominant source of water for in-stream flows, particularly during droughts and dry seasons. Even in relatively high-rainfall regions, groundwater extraction can have major impacts on stream flows. In England and Wales, for example, the former United Kingdom National River Authority "has drawn up a priority list of 40 locations where unacceptably low river flows are considered to be caused by excessive authorised abstractions, rather than drought" (Grey et al., 1995, p. 7). Along the coast of the western United States, salmon and other fish species are major contributors to regional economies. Some stocks are now threatened with extinction. While much of this is due to dams and changes in surface-water management, groundwater extraction may be a contributing factor. The San Joaquin River, for example, once supported significant salmon runs. Now long sections of the river are dry, because of a combination of groundwater extraction and upstream diversions.
- ***Coastal ecosystems.*** These ecosystems are often highly productive. Sea-water intrusion into coastal aquifers is a well-known cause of environmental damage to coastal ecosystems. Potential impacts on estuarine and marine ecosystems also exist. Productivity in these ecosystems is highly dependent on the balance between freshwater inflows from surface water, groundwater discharge and saline ocean water. Disruption of this balance through diminution of groundwater contributions to base flow could have major effects on the coastal environment and, through that, on the economic value of fisheries and other human uses.
- ***Inland wetlands.*** Wetlands are some of the most productive and biologically diverse inland ecosystems. In many if not most cases, water availability in wetlands depends on high groundwater levels. Wetlands generally have high biological diversity, support important fisheries (including spawning habitat) and are critical for waterfowl. In addition, they often absorb and reduce pollutants, assist with groundwater recharge and provide flood control benefits. In the United States, states have lost on average more than 48 per cent of their original wetland areas (Gleick, 1993, pp. 295-296). In California, wetland areas have declined from historical levels of 3 million to 5 million acres to perhaps 450,000 acres: a drop of 85-90 per cent (Department of Water Resources, 1994b).

Dry season base flows in the Ruaha Valley, Tanzania, sustain both aquatic and terrestrial biodiversity.

- ***Surface vegetation.*** Groundwater levels directly influence many vegetation communities. Phreatophytes, plants that derive a major portion of their water needs from saturated soils, can be the dominant vegetative species in ecosystems where groundwater levels are shallow. They often form critical wildlife habitat and may serve as important sources of food, fuel and timber. They also utilize substantial amounts of water. Reductions in evapotranspiration through removal of phreatophytes can reduce demand on groundwater resources, freeing water for other uses or causing water levels to rise and thereby leading to waterlogging and other environmental problems (see box 17).

In many areas, the environmental values dependent on groundwater conditions are closely intertwined with a broad array of human use patterns. Groundwater is an integral part of linked hydrologic, ecologic and human use systems. Changes in surface-water use, groundwater use or vegetation can send ripple effects throughout these interlinked systems—often effects that are difficult to predict. The degree of interlinkage is well illustrated by the case of the San Luis Valley in Colorado, where changes have the potential to affect deeply rooted cultures as well as specific environmental values.

2.3 Inherent externalities

Production of each groundwater-dependent service has inherent externalities, particularly where available resources are limited. For example, ensuring that rural populations have as much access to groundwater as possible in order to meet food security, poverty alleviation and rural development goals often implies allocation of that water away from other environmental or domestic water-supply services. In addition, production of each service generally involves some consumptive use, water quality generally declines and in many cases hydrologic system dynamics are altered. These changes generally alter the availability and quality of groundwater for other uses.

Another form of externality is the trade-off between efficiency and equity or poverty alleviation. From an economic perspective, it is often argued that groundwater should be used as efficiently as possible both within and between uses (National Research Council, 1997). Policies designed to ensure equitable access

Farms irrigated by groundwater on the outskirts of Ta'iz, Yemen, impose drawdown externalities on down-gradient users.

to groundwater as a "common heritage" or to reduce poverty by ensuring that marginal farmers can afford to pump groundwater generally come at the expense of economic efficiency and may have major impacts on other groundwater services. The case of India is illustrative (World Bank, 1998a). There, high subsidies for both installation of wells and power for pumping have played a major role in ensuring that rural farmers have access to groundwater. As a result of these subsidies, the number of energized agricultural wells in India has increased exponentially from about 3,000 in the late 1940s to over 20 million at present. Across large areas, even small farmers now own wells and have been able to increase food production greatly as well as raise their own standards of living. Inefficient patterns of water use and growing groundwater overdraught problems have, however, emerged as major externalities associated with these policies. Low prices for pumping energy provide farmers with little incentive to avoid waste, and increased access to groundwater has been a major factor contributing to the widespread emergence of overdraught concerns.

The vulnerability of aquifer systems has spurred significant concern for the environmental impacts of groundwater resource depletion, quality and pollution (Robins, 1998). In developed countries this has led to a shift away from new groundwater development to groundwater management and protection, leading in turn to a much greater emphasis on understanding groundwater processes. This may have come too late for many aquifers in the already industrialized countries and has led to widespread and expensive groundwater clean-up in the United States and Europe. Nonetheless, the hydro-environmental imperatives faced by many developing countries highlight the need to put in place management approaches that are realistic about maximizing access to safe groundwater but which at the same time regulate the development in a supportive fashion on the basis of a sound appreciation of hydrogeological processes, limits and risks. In the rift settings of west and east Africa, many large-scale rural drilling programmes have subsequently encountered increasing fluoride content in shallow groundwater. The wells needed to be drilled, but geochemical risks in the basement complex and mixing of geothermal groundwater were not properly evaluated when the drilling campaigns were planned (Smedley et al., 1995). Similar problems related to arsenic are now widespread in South Asia.

The core point to recognize here is that each of the services produced by groundwater also has inherent externalities. These externalities are particularly evident where aquifers are vulnerable to pollution or quality changes and where groundwater is limited and subject to competition (see box 18).

3 View to a future of increased competition

The groundwater-dependent social and environmental services discussed in the preceding chapter are important to different groups of users and stakeholders. As subsequent sections will document, these services are in many cases threatened by overdraught and pollution. Overall, the increasing demands of society place an ever greater burden on the groundwater resource base. This has created a fundamental change in the structure of social dialogue over groundwater resources. In the past, groundwater had been developed with little thought for other users, but is it is now apparent that competition for groundwater services is growing. Disputes over access to groundwater and allocation of the social, financial, technical and other resources necessary to manage the resource are also intensifying. Finally, competition is intense over the underlying social objectives management should address, the mechanisms through which management should be initiated practically and the political philosophies underlying management approaches.

Competition is likely to increase substantially over coming decades. World population is projected to approach 8 billion by the year 2020 (United Nations, 1998a). It is growing at 1.33 per cent per year (2.04 per cent in the period 1965-1970). The medium-fertility projection, usually considered as "most likely", indicates a world population of 7.5 billion in 2020 and 8.9 billion in 2050; 97 per cent of this increase is occurring in the developing regions of the world. Perhaps more importantly, the distribution of the population is also changing. Less than 15 per cent of the population lived in urban areas in 1920; this had increased to 30 per cent by 1950 and is projected to increase to 61 per cent by 2030 (United Nations, 1996). In coming decades, urban populations will exceed those living in rural areas. Projections indicate that by 2030 86 per cent of the growth in urban populations will have occurred in developing countries (United Nations, 1996a). The implications for the patterns of groundwater use and resulting subsidence, ingress of low-quality water and pollution in urba areas are manifold (Chilton et al., 1997).

Urbanization causes fundamental changes in the structure of demand for water. Urban centres are generally centres of political power. Populations residing in them demand access to high-quality, reliable supplies for domestic uses. Furthermore, urban populations generally depend on industrial and other non-agricultural activities for their livelihood. These industrial and commercial portions of the world economy are growing rapidly. As a result, water demands associated with these activities often increase, backed by the political and economic power of

Figure 5 Percentage of population residing in urban areas, 1950-2030

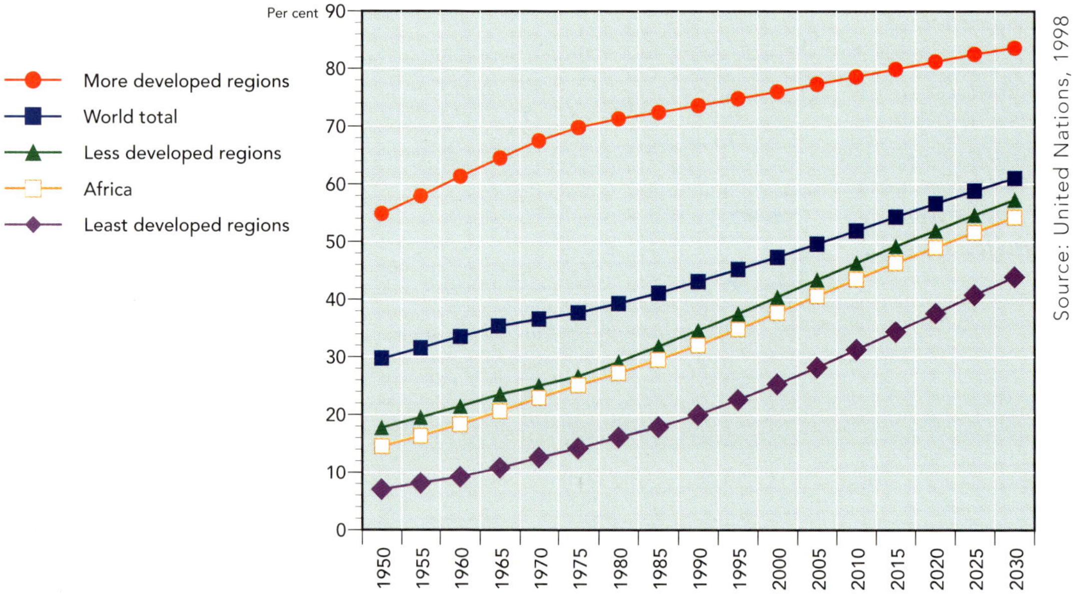

growing urban populations.[5] Shifts in political and economic power, while reducing the political and economic influence of farmers and rural populations, will not reduce the need for food production. A population approaching 8 billion requires massive amounts of food. Irrigation has been the lead input catalysing yield increases over the last five decades. Even if local and national efficiency in food production can be improved, global irrigation water demands are unlikely to decline greatly.[6] Finally, environmental water requirements are increasingly recognized. Over the last two decades, political support for environmental maintenance and allocation of water to meet these requirements has grown in many post-industrial countries. Much of this support has come from urban populations seeking to maintain the amenity values of rural areas.

---

[5] Increased dependence on commercial and industrial activities does not inherently require increased water withdrawals for those sectors. Excluding water used for thermoelectric generation, withdrawals of water for industrial uses in the United States, for example, declined in 1995 nearly 40 per cent from their peak in 1970. Industrial water withdrawals during 1995 in the United States were, in fact, over 20 per cent lower than in 1950 despite massive growth in industrial production. This decline has occurred primarily through conservation measures and increased use of recycled water (which grew by 36 per cent between 1990 and 1995). In contrast, despite declines in per capita water withdrawals for domestic use (public supply), the amount of water consumed has increased steadily over the same period as the population has grown (Solley et al., 1998).

[6] Withdrawals of water for irrigation in the United States, for example, declined 2 per cent between 1990 and 1995 and over 20 per cent between 1975 and 1995. Declines in withdrawals, however, are not equivalent to declines in use. Consumptive uses of water in the United States for all uses increased 6 per cent between 1990 and 1995 despite an overall decrease in withdrawals of 2 per cent (Solley et al., 1998).

The growth of urban centres such as Ta'iz, Yemen, is placing unreprecedented levels of demand upon aquifer systems.

At the global level, demands for both bulk and retail water supplies and the raw water resources base itself can be expected to increase over the next two decades, accompanied by shifts in the political and economic structure of the demands. In this rapidly changing environment, competition over available groundwater supplies will increase along with that for surface-water supplies. With the exception of specific cultural preferences, the consumer is largely indifferent to the provenance of the water. Therefore, a core challenge for water managers in coming decades will be to ensure that competition does not escalate and deadlock society's ability to address groundwater management needs. Meeting this challenge requires recognition that competition is not simply a fight over access to the groundwater resource base, but has multiple dimensions.

3.1 Dimensions of competition

Four core forms of competition are important to recognize in the context of groundwater: among users, among uses, among regions and among philosophies. These different dimensions each have fundamental implications for approaches to managing groundwater resources.

3.1.1 Among users

In areas where water-level declines and/or pollution limit the availability of groundwater, competition among individual users over access to the resource can be intense. This often takes the form of competitive well deepening, although there can be a trade-off with water quality which can deteriorate with depth. In addition, in many cases less wealthy farmers are unable to maintain access to groundwater when competition occurs. This can result in farmers returning to dryland agriculture and even migration to urban areas.

Competition is not just a rural phenomenon. Competition for scarce water in urban areas often has major equity and health implications. In most cities in developing countries, large portions of the population, particularly the poor, are not served by municipal supply or sewage systems and depend on shallow wells as a primary source for domestic water needs. Water-level declines frequently cause

Multiple pump owners accessing a single well in Rajasthan

In the urban setting, the nature of competition is essentially economic, rather than physical

shallow wells to go dry. This leads to water scarcity, dependence on limited numbers of municipal tap stands or the necessity of purchasing water. In the first case, people—particularly women—are often forced to wait for many hours to obtain water. Water purchases, on the other hand, can represent a major expense. In urban areas in some countries, for example in Yemen, limitations on access to water through municipal systems and traditional shallow wells have caused the emergence of water markets. Water is sold to residents by private suppliers who own deep private wells or purchase water from farmers and transport it by tanker into the city. In the city of Ta'iz in Yemen, stratified water markets have emerged

Private water tankers accessing private boreholes on the outskirts of Ta'iz, Yemen

in which price varies with water quality. Low-quality water is sold for washing and cleaning uses, higher-quality water for cooking and purified water (supplied from small, private treatment plants) for drinking. At the highest level, purified drinking water can cost as much as Yemeni rials 2000/m^3, while lower-quality water for washing can cost as little as YRls 150/m^3.[7] These prices compare to daily wage rates for labourers of YRls 150-200 in 1995. While drinking water is provided free of charge at locations such as mosques and from many private wells, the price of retail water on the open market can be prohibitive for the urban poor.

Competition among users is a local phenomenon, but the social dynamic it creates has regional implications. If numerous farmers who are unable to afford the cost of deepening their wells (where deeper aquifers are available) lose access to groundwater, competition could become a force driving agricultural consolidation

[7] Costs in 1995/96. At that time the exchange rate varied between roughly 85 and 125 Yemeni rials/United States dollar on a daily basis.

Box 7 Urban-rural competition for groundwater: the case of Ta'iz and Al-Hima, Yemen

Throughout Yemen, urban areas and their associated water demands are growing rapidly. In the city of Ta'iz, these demands have been provided by NWSA (National Water Supply Authority) primarily from the Al-Hima and more recently the Habir areas. These sources are insufficient, however, and during some periods water-supply frequency in Ta'iz has declined to 45 days. Consumers meet the balance of their needs by purchasing tanker loads of water at high cost from private suppliers. Water-supply problems in Ta'iz are not new. Emergency deep-well-drilling programmes first in the Al-Hima and later in the Habir areas were undertaken in 1989/90, 1992 and again in 1995. Wells drilled during these emergency efforts tapped deep volcanics and the Tawilah sandstone to supplement supplies available from shallow wells in upper alluvial wadi bed aquifers. Unfortunately, yields declined rapidly. Well field production in 1993 following the 1992 emergency programme totalled 4.59 million cubic metres (Mm^3); by 1995, this had declined to 2.6 Mm^3. Declines in production are not confined to municipal wells. Many wells belonging to local farmers have failed. As a result, tensions over water transfers from Al-Hima and Habir to Ta'iz have increased dramatically—including, in some instances, armed conflict over new municipal wells.

The irrigated wadis near Ta'iz look like green bowls. Prosperous-looking farmhouses dot the fields. The impression is misleading. Drops in the water-table are not visible at the surface. Crops remain green and areas prosperous—until the wells run dry. In Al-Hima following the emergency programmes to supply water to Ta'iz, water levels have declined substantially and lower areas along the wadi have dried out completely. In the 1970s and early 1980s, lower Al-Hima was a vibrant agricultural community. Now, as the photograph illustrates, piped water supply vies with newly planted qat fields and most agriculture now depends on rainfall. Drought-resistant millet, which produces only a small crop of grain and fodder, has replaced the high-value fruit, vegetable and qat crops that provided the economic base for local villages. Even drinking water is in very short supply. Children, women and men travel long distances by donkey or camel to collect water at the few tap stands that still run.

With the decline in agriculture, populations in the lower Al-Hima area have been forced to depend on other activities to support themselves. Many families survive from hand to mouth. Income from a brother, son or father working abroad or in the city is the primary basis for survival. Most of the men remaining in the village travel out to Ta'iz city daily and seek work as casual labourers. Poverty has become a way of life and few see avenues to improve their condition.

Habir, upstream from Al-Hima, views itself as the next victim of development to supply Ta'iz. Drilling in Habir began in April 1995 and was initially disrupted by local inhabitants using petrol torches. A series of violent incidents followed, during which many local leaders were imprisoned and two women were shot and seriously injured by soldiers attempting to enforce the drilling. Compensation for water transfers from Habir to Ta'iz has been agreed—but never delivered. Tensions and distrust are high.

Drilling a well for municipal supply near Tai'iz, Yemen

and migration to urban areas. This may in turn result in intensive competition between peri-urban and centralized systems tapping the same aquifer (see box 7). This is, for example, the case in many cities in India, where private home owners install wells into aquifers that serve as a major portion of the supply for the urban area. It is also the case in Lusaka, Zambia, where high-income suburban dwellers will pay for private boreholes on their own properties (in which they have user rights over groundwater). This is in preference to poorly treated and unreliable municipal supplies pumped from the Kafue River, some 40 km to the south of the city. The assurance of quality and reliability may be more easily found in the local dolomite aquifer system but it will prove sustainable only if withdrawals are regulated and the aquifer is protected from pollution. In addition, expansion of low-income areas on the same aquifer system with no sanitation combined with pollution from industrial zones has created a large pollution risk. This situation is common in many countries where investments in water supply and sanitation have not kept pace with increasing urban populations. The total absence of groundwater regulation above a prolific but sensitive aquifer system may eventually result in the total loss of the groundwater resource asset.

3.1.2 Between sectors and uses

As with individual users, scarcity results in competition among uses. As a starting point it is essential to recognize the distinction between high-volume uses, such as agriculture, which can often tolerate wide variations in water quality, and low-volume domestic uses, which generally require high-quality supplies. This distinction can place a particular premium on groundwater supplies.

Worldwide, agriculture is the largest single user of groundwater. Although good-quality data on groundwater extraction are sparse, most estimates suggest that 80 to 90 per cent of global groundwater extraction is dedicated to irrigated agriculture. Pressures for water to meet the domestic, environmental, industrial and other "high value" demands have led to increased competition between these sectors and agricultural users. In California, for example, analysts have called specifically for reductions in agricultural water consumption in order to address overdraught concerns and meet the demands of other users (Gleick et al., 1995). In many regions where surface-water or groundwater supplies are limited, such reallocations from agriculture are seen as the primary source of water to meet other needs. The logic behind the demand for reallocation of water from agriculture is strong. In many situations, agriculture uses high-quality groundwater and the reallocation of even a small fraction of that supply to urban uses would not have a large impact on overall agricultural production. Furthermore, in many cases reductions in agricultural water-use from high-quality sources could be offset by increasing water use efficiency and switching to available lower-quality water sources.

Regardless of the potential benefits of water reallocation, competition over groundwater is focused and intense. Farmers owning land adjacent to growing urban centres will often oppose the construction of urban water-supply wells within their area. In the western United States, the history of Owens Valley hangs over most water reallocation debates. The Owens Valley was once a rich agricultural area. Rights to both surface water and groundwater in the Owens Valley were purchased by the growing city of Los Angeles and the water was reallocated from agriculture to urban uses. Now agriculture in the valley has disappeared and a once-prosperous area is mostly desert (Reisner, 1986). This case highlights one of the reasons why competition is so intense. While reallocation of water from agriculture to other uses might have a minimal impact on the agricultural sector as a whole, within local regions the impacts can be tremendous. Furthermore, agricultural water users are critically dependent on the timing and reliability of water supplies. When periods of agricultural water needs (for example, the germination and flowering stages of many crops) coincide with periods of high demand by other users, availability constraints can have much larger impacts than suggested by aggregate water-use figures within sectors or economic returns per cubic metre of water consumed. As a result, analyses of potential gains from water reallocation must be critically evaluated with respect both to the factors considered when the analysis was undertaken and to the dynamics inherent in specific situations.

The *in situ* value of groundwater is another point where competition between uses has emerged. As previously noted, in many cases environmental values, such as stream flows, are affected by groundwater levels. Maintaining these values may require leaving large amounts of groundwater in aquifers to maintain specific piezometric and phreatic water levels. Similarly, maintenance of a buffer against drought or climatic fluctuations depends on reserving groundwater during wet or normal precipitation years for use during periods of scarcity. These *in situ* values and the uses dependent on them often conflict with other use values and allocation mechanisms. In India, for example, the objective behind government subsidized well-financing schemes has been to increase irrigation well numbers until the "sustained yield" of aquifers is reached—assuming that this "sustained yield" is somehow "known". The goal of maximizing irrigated area for food production ignores the likelihood that full extraction may have a number of negative consequences, such as: (1) reduction in groundwater availability during droughts; (2) impacts on surface-water sources; (3) water levels falling below the depth of shallow wells; (4) mobilization of low-quality water sources; and (5) aquifer pollution. These consequences may, in turn, conflict with other basic objectives, such as poverty alleviation or environmental maintenance.

A key factor to note in the above discussion is that competition between uses is not limited simply to groundwater access. *In many ways, the sharpest points of competition between uses may have to do with the objectives of management, and not necessarily with allocation of the volumes of water available.* Many uses, such as provision of access to drinking water and maintenance of base flows, may not require substantial amounts of water in themselves. Instead, they depend on maintenance of specific aquifer conditions, water levels and water quality. Maintenance of these conditions may restrict other uses long before there is any limit to the actual volume of good-quality water available.

3.1.3 Between regions and nations

Often, competition over groundwater between uses means competition between regions. In the western United States, water reallocation from agriculture to distant urban areas has led to a major debate over impacts on areas of origin (Moench, 1991). This type of reallocation can cause ripple effects throughout agricultural economies as agricultural services, the jobs they support and tax revenues decline. As a result, reallocation of water is often perceived by local residents as undermining future economic possibilities for entire regions. Debates over reallocation are, as a result, intense. In some cases, such as debates over management of the mountain aquifer underlying Jerusalem (see box 8) and groundwater pumping along the United States–Mexico border, changes in groundwater flows between regions can have major international implications for the future of nations.

Competition between regions over scarce water supplies is, however, not always negative. In instances such as the Israeli-Palestinian debate over the mountain aquifer outlined in box 8, competition can become a basis for dialogue and the resolution of joint issues. It can, thus, serve as a starting point for mutual understanding and ultimately management systems. Competition can, however, also result in deadlock. Tendencies in that direction may be most common where competition over the resource itself brings more deeply held philosophical differences to the surface.

3.1.4 Between philosophies

Value systems and their underlying philosophies can affect groundwater management. Put simply, some argue that water must be recognized as an economic good and allocated to uses with the highest economic returns. Others view water primarily as a common heritage to which all people, and often the natural environment as well, have fundamental and inalienable rights. The precise philosophical underpinnings of these arguments are not always clearly defined or appreciated (Livingston, 1993; Waldron, 1994). The common-property nature of aquifer systems and their inherent vulnerability make groundwater particularly sensitive to these polarized views. Competition between such philosophical viewpoints is often sharpened by different conceptions of the role of governance and the political process (see box 9).

In Islamic societies, individuals provide wells as a public service. An example from Tunisia.

Similar fundamental philosophical conflicts are inherent in different approaches to managing groundwater (Barraqué, 1997b). Some argue for State ownership and control over water resources as a reflection of the "common heritage" nature of the resource. This frequently leads to prescriptions for strong centralized control and regulation. Others argue, based on views that contrast "the State" and "the people", that management must be through local institutions using democratic governance structures. Still others focus on the sanctity of private rights and the efficiency of markets as allocation mechanisms. Philosophy may, however, conflict with pragmatic recognition of what will work in different cultural and physical contexts. Competition between philosophical perspectives is often unrecognized and greatly complicates the development and replication of avenues for addressing groundwater problems.

Box 8 Palestinian and Israeli debates over the mountain aquifer

Palestine and Israel share many resources, but especially water. The most important of these water resources is the so-called "mountain aquifer". This aquifer is mainly limestone. Flow through it is rapid and the geology is highly complex. Approximately 90 per cent of the catchment is under the West Bank Palestinian territory and 60-70 per cent of the storage is under Israel's pre-1967 borders. The result is an aquifer that would be a political problem if it underlay the boundary of New York and New Jersey or the United States and Canada. In Israel and Palestine, the situation is that much worse because of a history that is evident to everyone, but that was complicated by the isolation of researchers from one another in the years after 1967.

Israelis and Palestinians from a range of disciplines—law, economics, hydrology—proposed a joint study of joint management in 1992, almost a year before the peace accord. Together with the CRB Foundation, IDRC (International Development Research Centre) funded the project. The research effort is a classic example of why aquifer analysis must be approached as a process, particularly where management is concerned. Once the peace process started, this work was construed as academic or third-track activity complementing more formal political-bilateral and technical-multilateral tracks. Some of the people who participated in diplomatic negotiations, especially on the Palestinian side, participated as analysts in this study.

Almost from the start, two options for management were rejected: (1) separation of management activities, because it is physically impossible; and (2) domination by one side, because it is ethically and politically unacceptable. Therefore, a joint management approach was selected—but what this would entail and how it might work remained to be defined. As joint management began to be investigated, some things were learned early: (1) tough issues, such as quantity allocation and water rights, could be put to one side for the moment; (2) technical management of the aquifer could be delayed—it will be required in any case; and (3) no existing management model was adequate.

Early in their work, the research teams defined four basic structures for joint management depending upon the function of the agency involved: (1) resource protection—to avoid loss of water quality; (2) crisis management—to respond to both crisis (such as a spill) and drought; (3) economic efficiency—to emulate the results that would come with a private market; and (4) integrated aquifer management, possibly with regulatory powers. The main body of the team decided that these partially overlapping models should be developed sequentially, starting with the simpler issues and moving towards those that require higher degrees of cooperation. However, at some of the workshops, a minority view was expressed to the effect that, at times, history allows you to go for the whole thing, and that the team should not resist recommending higher degrees of integration.

Adopting a sequential approach, all structures included some elements of the following: data collection and monitoring of flows and quality, research, short-term supply augmentation, and a formal system for conflict resolution. A blueprint was not provided because it was neither appropriate nor possible to do so. Not only was the research an example of learning by doing, but it would also be the actual management process. In addition, alternative models of public participation were discussed but not included in any formal structure. With this as a base, each of the four structures was modelled qualitatively. The result was a matrix of 19 functions that might be filled by an agency, with each function described according to staffing levels, funding requirements, and degree of cooperation required etc.

As the study progressed, the teams found that often the most controversial issues were sectoral, not national. If Israeli farmers suffer, so too will Palestinian farmers. It also proved easier to respond to quality issues rather than quantity allocation. Indeed, quality issues stimulated cooperation. Finally, it is much more productive to focus on the process of creating a management structure than on specific end goals.

Box 9 Competing philosophies

"By the law of nature these things are common to mankind—the air, running water, the sea and consequently the shores of the sea."

Institutes of Justinian, 2.1.1

"The proposed new approach to managing water resources builds on the lessons of experience. At its core is the adoption of a comprehensive policy framework and the treatment of water as an economic good, combined with decentralized management and delivery structures, greater reliance on pricing, and fuller participation by stakeholders."

World Bank, 1993, p. 10

Groundwater, like other natural resources, is widely viewed as a common heritage. This perspective often conflicts with its role as a foundation for economic activity and the increasingly expressed importance of treating it as an economic good. Those who emphasize water's role as an economic good generally also advocate the development of private rights systems—often in conjunction with water markets—as social mechanisms to encourage use efficiency and prevent over-exploitation. Approaches based on the establishment of private rights conflict directly with traditions that recognize water as a "gift of God" or inalienable component in the natural heritage of citizens within a country.

In the United States, the public trust doctrine has been used effectively to limit the ability of private right holders to transfer water from rural areas where it played a major environmental role to urban uses in the city of Los Angeles.Writing on this case, Khoeler (1995, p. 548) argues that: "As an attribute of sovereignty, the public trust cannot be shaken off by the state through legislative abolition or even through Constitutional prohibition. The California Supreme Court has determined that the public trust embodies the state's duty to protect the 'people's common heritage' in natural resources." Similarly, in Islamic tradition, Mohamed's saying that "People are partners in three [things]: fire, water and grass" emphasizes the common nature of water rights and has been interpreted by some scholars as forbidding water sale, whether for drinking or other uses (Al-Eryani, 1995). The shariah, or "way", originally connoted the "path to water" (Wescoat, 1995, p. 5). It provides the ultimate basis for "rights of thirst" that apply to both humans and animals and that extend throughout the main Islamic systems of jurisprudence. Water is a gift of Allah, and a broad set of social duties within Islam forbid refusal of water for human or animal needs, particularly surplus water. This applies whether the source is groundwater or surface water (Wescoat, 1995, p. 11). In India, deeply rooted traditions place similar emphasis on the common nature of water resources. As Singh notes, space, air, water and energy have traditionally been viewed as "incapable of being bound into property relations. No dharmasastra or vyavahara text mentions property rights of anyone, including the king, in rivers or streams" (Singh, 1991, p. 26).

Conflicts between perspectives rooted in tradition and religion and perspectives emphasizing the economic role of water are of critical practical importance. In many countries, these conflicts are a fundamental factor blocking water law reform or the establishment of other frameworks essential for management. Furthermore, even if legal reforms are passed and management frameworks are formally established, philosophical debates often remain as major points of social tension and can limit a society's ability to take effective action.

3.2

Prospects for the future

If groundwater problems and competition remain unresolved, the social and environmental consequences and their associated costs will be significant. Emerging problems directly undermine the sustainability of key social and environmental services society depends upon. Competition, on the other hand, complicates society's ability to address problems and, in extreme cases, can lead to disputes over groundwater allocation and access.[8] Competition, however, has positive sides. When common goals can be identified, the process of resolving competition over water resources can provide a basis for dialogue and understanding between widely different or opposing sections of society.

Perhaps the most fundamental issue facing those concerned with the sustainable development and management of water resources is ensuring that competition and conflicting perspectives do not result in deadlock. Competition among users, uses and regions can often be resolved when different groups understand the issues and needs that others face. Competition between philosophies is more

Farmers discussing water allocation with officials in Haryana State, India

difficult to resolve since it often reflects fundamentally different world-views and the legacy of past administrative approaches, be they rooted in common law or civil code. As Barraqué (1997, p. 1) notes in his review of European approaches to groundwater management, "Some American and Spanish hydrogeologists have even argued that the low visibility of groundwater had led water engineers to practice a sort of 'hydroschizophrenia' between both types of resources. In the opinion of policy scientists and lawyers, however, the reason for this is not solely the hidden and slow-moving character of aquifers: it is also rooted in their private status, linked to the belief that they were plentiful wherever they could be found."

---

[8] Examples of actual conflict include the situation near Ta'iz in Yemen, Jodhpur in Rajasthan and the Owens Valley in California.

In seeking to address competition, it is important for policy makers to recognize the practical power that individual well owners wield over groundwater conditions and the limitations of governmental power over actions by those individual well owners. Wells and boreholes are generally located on private lands and use can be difficult to monitor. While centralized control over groundwater pumping may be practical in some situations, such as in municipal well fields or irrigation schemes, in most situations it is not effective without the support of the water users. As a result, points of competition often cannot be resolved through centralized water allocation or management decisions. Instead, conflict resolution frequently requires the growth of understanding between users. The process of dialogue described in the Israeli-Palestinian case points a way forward towards the generation of mutual understanding (see box 8).

What the future may hold depends heavily on whether or not processes capable of generating mutual understanding of management needs and mechanisms can be established. Such processes must address both deeply held philosophical differences and issues of access to resources that are critical to the economic survival of many populations. If well owners, those who often have the greatest de facto control over groundwater use, lack ownership of the process, the mutual understanding essential for allocation and management decisions to translate from theory into reality will not exist. Resolution is likely only if the processes are free and open. Furthermore, processes must address the root causes of competition—not just the immediate factors sparking its emergence in any specific situation. For these reasons, it is essential to examine both the proximate and root causes of groundwater problems.

The next chapter provides a brief overview of specific groundwater problems (overdraught, pollution, waterlogging and the migration of low-quality waters) and their proximate causes. Following this, the underlying social factors that members of the ad hoc group view as contributing to the emergence of current problems are discussed. This discussion represents the foundation we have used to generate insights into ways forward—the basic principles important to addressing degradation of the groundwater resource base.

4 Emerging threats to the resource base

The multiple roles groundwater plays in meeting the needs of society and the natural environment are threatened in many areas. In some cases, rapid increases in extraction over recent decades have led to rapidly falling water levels and irreversible damage to aquifers. As extraction grows, groundwater flow patterns change, drawing low-quality water into fresh aquifers. Pollution is also increasing. Point-source discharges from industries and urban areas create intense, but generally localized, pollution problems. The gradual pollution of aquifers from non-point sources such as agriculture represents a potentially more insidious threat to groundwater resources. Worldwide increases in food production over the last five decades have been driven by an irrigation-led package of technologies dependent on the use of chemical fertilizers and pesticides. Return flows from agriculture

Pollution is not always obvious. The low-intensity, non-point sources of pollution from irrigation return flows may look benign, but can result in lasting damage to aquifers.

are leading to gradual increases in nitrates, heavy metals and other contaminants. Although the impacts of dispersed pollution are difficult to monitor and quantify, they clearly add a significant burden to the environment and human health. Furthermore, once aquifers are polluted, remediation is often impossible or uneconomic. This is because pollutants are subject to various physical processes once

Box 10 Over-abstraction and sustainability: complex theory, simple practice

There remains a confusion in the usage of "over-abstraction" and "groundwater mining". The latter only refers to the depletion of a stock of non-renewable groundwater that will not be replaced, leaving the aquifer de-watered indefinitely. Clearly, the planned mining of an aquifer is a strategic water resource management option if the full physical, social and economic implications are understood and accounted for over time. However, the replenishment by downward percolation of meteoric water shows high inter-annual variability and is a complex physical process that is difficult to evaluate (Lerner, 1990; Simmers et al., 1992). Therefore, over-abstraction should not be defined in terms of an annual balance of recharge and abstraction, but needs to be evaluated on an inter-annual basis, since the limit between the non-renewable stock and the stock that is replenished by contemporary recharge from surface percolation is usually unknown.

What really matters to decision makers and well users is the overall reliability and productivity of a well (in terms of water levels, volumes and water quality) during a given time period. Therefore, if a well taps a particular aquifer, what is its sustainable rate of exploitation given variable periods of recharge and drought?
The answer to this question is not trivial, and requires a certain level of precision in understanding the system dynamics. Since most aquifer systems are complex (heterogeneous and anisotropic, leaky and bounded), they respond to pumping or injection under non-steady-state conditions (drawdown stabilization is a very rare exception). Drawdowns in pumped wells are the result of two components: (i) a time-dependent portion related to aquifer properties and prevailing boundary conditions and (ii) a well loss component that varies linearly and non-linearly with the pumping rate and is independent of time. Under these circumstances, it is possible to translate socio-economic and technical criteria (including pumping costs, mobilization of low-quality water and saline intrusion) into a value of the maximum available drawdown (MAD). The actual sustainable exploitation rate (SER) of a pumped well is then related to the MAD for a given period of time, *t* (usually in years), as follows:

$$MAD=\{(SER/4\pi T)f(S,T,t,r^2)+A(SER)+B(SER)^a\}$$

Where T is transmissivity, S is storativity of the aquifer, r is the radial distance to the well, t is time and A, B and a are well-loss coefficients. The function f integrates the aquifer flow and boundary conditions, including partial penetration interference effects.

To resolve this non-linear equation, numerical models are required when analytical solutions are not available (Sauveplane, 1987). This method provides values of the sustainable exploitation rate of a given well over a given time period. Given a set value for MAD, that could be, for example, 75 per cent of the depth between the static water level in the pumped well and the level of the pump. More sophisticated criteria can be employed, for example when an economic model is coupled with a physical model to manage optimal exploitation according to the set SER criteria. Over-exploitation of the well can be precluded by assigning SER as the maximum admissible pumping rate of the well.

In the case of a well field, interference effects need to be evaluated to assign individual SERs and a combined SER for the field. Predictions of drawdowns at observation wells for fixed SERs, *t* and analytical solutions are calibrated with observation well data. Progressively, this method can be extended over a whole aquifer system, with global values of MAD assigned to avoid over-exploitation. If the aquifer system is sufficiently well known, the assigned value of MAD may also include the exploitation of a portion of the non-renewable groundwater resources.

Such methods can provide a basis for pre-empting aquifer system degradation before physical and socio-economic damage is done, by giving indications to users of sustainable exploitation rates.

introduced into the aquifers (dispersion, diffusion and sorption on mineral particles). Their chemical nature can also be changed and as a result they often persist even when the pollution source is removed. Pollution can, thus, eliminate aquifers as water sources as effectively as physical depletion of available groundwater supplies. Groundwater can also be tainted with harmful concentrations of naturally occurring elements, such as fluoride and arsenic—or deficient with respect to certain elements critical to metabolism—iodine, for instance. These geochemical threats to human and animal health are in fact pervasive (Appleton et al., 1996) and are particularly acute in the case of naturally tainted or mineralized groundwaters, where abstraction can lead to excessive concentrations (Edmunds and Smedley, 1996).

Addressing the above concerns represents a major challenge for society over the coming decades. The ubiquity and invisible nature of groundwater resources, combined with the gradual nature of many emerging problems, make these problems easy to ignore until they are beyond practical solution. At the same time, there is little point in investing massive resources in attempts to solve all perceived problems rapidly. Understanding both the physical dimensions and the underlying social causes of emerging problems is essential. Building the basis for management will also require a long-term process of management capacity development. But it is important not to overly dramatize current trends. In many regions, productive aquifer systems are large and will respond slowly both to threats and to management so that long-term efforts will need to be sustained and not crowded out by brief bursts of attention.

Hence it is essential to outline, as clearly as possible, the nature of emerging threats to the resource base, perspectives on their implications for the future and their underlying causes if sustainable management responses are to be implemented.

4.1 Over-abstraction and water-level declines

The abstraction of groundwater from aquifer systems has accelerated rapidly in the twentieth century after the introduction of motorized pumps. This has led to the notion of "safe yield" or "sustained yield" for pumped wells and and has been part of the groundwater engineering vocabulary throughout the twentieth century (Meinzer, 1920; American Society of Civil Engineers, 1961; Miles and Chambet, 1995). However, this concept is limited—it only considers renewable groundwater and ignores vertical flows (see box 10). Indeed, it was only in the last decade of the twentieth century that the scale of economic and social impacts related to groundwater use started to be addressed in professional forums. An International Association of Hydrogeologists Congress and a UN conference, both dealing with over-exploitation, were held in the Canary Islands in April 1991 (Simmers et al., 1992; United Nations, 1991) and provided key papers on environmental, economic, policy and legal issues. More recently, further work has been published dealing with questions of sustainability and the vulnerability of aquifer systems (Robins, 1998). In many instances this points to the appropriateness of source vulnerability indicators rather than state-response indicators when considering impacts of over-abstraction and pollution on groundwater.

Current levels of groundwater abstraction over and above the natural rates of replenishment are already significant, but aquifer systems exhibit a variety of responses to stress that require some judgement before alarms are sounded. This observation notwithstanding, examples of over-abstraction are well documented.

The Ogallala aquifer, which underlies eight states of the midwestern United States, represents a classic case of overdraught. Water-level declines in some areas have exceeded 35 metres since 1950 because of extraction rates that are far in excess of recharge (United Nations, 1995). In Rajasthan, India, maps prepared by the state groundwater department show water-level declines between 1984 and 1994 from 3 m to more than 10 m in a broad belt across the central portion of the state. This trend has continued since 1994 despite successive years of above-average precipitation. In some areas, declines greater than 2 m per year are common.

Water-level declines and groundwater overdraught can lead to a wide array of social, economic and environmental consequences, including:

- Critical changes in patterns of groundwater flow to and from adjacent aquifer systems;
- Declines in stream base flows, wetlands etc. with consequent damage to ecosystems and downstream users;
- Increased pumping costs and energy usage;
- Land subsidence and damage to surface infrastructure;
- Reduction in access to water for drinking, irrigation and other uses, particularly for the poor;
- Increases in the vulnerability of agriculture (and, by implication, food security) and other uses to climate change or natural climatic fluctuations as the economically accessible buffer stock of groundwater declines.

Yemen presents particularly severe evidence of the consequences of over-abstraction. According to the recent Water Resource Assessment of Yemen: "almost all important groundwater systems in Yemen are being exploited at alarming rates. Worst-case predictions made in 1985 on possible depletion of the Wajid sandstone aquifer of the Sadah Plain have unfortunately come true and groundwater levels have declined on average some 40 metres in only nine years." (Water Resources Assessment of Yemen, 1995, p. 104.) High-quality groundwater available in

Intensive irrigation in the plains north of Sadah, Yemen

Box 11 Arsenic in groundwater

Arsenic is a highly toxic element, and chronic ingestion can lead to serious health problems, most notably skin disorders (keratosis, pigmentation changes, skin cancer) and internal cancers. Most drinking water has low concentrations and represents a minimal risk. However, some supplies exceed the WHO provisional recommended limit of 10 micrograms per litre and can be a significant source of ingested arsenic. The recently recommended revision of the WHO limit and the greater awareness of potential arsenic problems have prompted closer examination of groundwater arsenic concentrations worldwide.

Groundwaters are particularly vulnerable to a build-up of arsenic because of their interaction with arsenic-bearing aquifers. Under certain conditions, the mobility of arsenic in solution can increase greatly. Although a natural phenomenon, the problem may be exacerbated by man's activity. Mining of arsenic-rich sulphide ores is a specific example of this. The ore minerals oxidize following excavation and release arsenic into water supplies. In such cases, arsenic is often not the only toxic trace element to be released.

Groundwaters particularly at risk from arsenic are those from aquifers containing sulphide minerals, iron oxides or acidic volcanic rocks. Arsenic is mobilized preferentially under reducing (anaerobic) conditions, but oxidizing groundwaters with high pH and alkalinity are also vulnerable. Some well-documented cases of naturally occurring high-arsenic groundwaters are found in Bangladesh, West Bengal, southwestern Taiwan and northern China, all of which occur in reducing sedimentary aquifers. Problems also occur in volcanogenic aquifers in Chile, Argentina and Mexico. In all of these affected regions, concentrations of arsenic in excess of 1 milligram per litre have been recorded and populations often exhibit serious health problems as a result of use of such water for potable supply.

Probably the most extensive case of arsenic poisoning from groundwater is that of Bangladesh and West Bengal. Arsenic occurs naturally in the groundwaters abstracted from the alluvial-deltaic sediments of the Ganges-Brahmaputra-Meghna river systems, and an area around 75,000 km^2 is thought to be affected by groundwater with high arsenic concentrations. As groundwater represents a major source of drinking water, many millions of people are potentially at risk from arsenic. In Bangladesh alone, some 20 million to 30 million people (British Geological Survey, 1998) are thought to be using drinking water in excess of the Bangladesh standard of 50 micrograms per litre and many thousands already exhibit arsenic-related health problems.

The arsenic problem in Bangladesh and West Bengal results from a combination of factors which lead to the release of arsenic from aquifer minerals and its retention in the groundwater. The arsenic is thought to be present mainly as easily soluble forms on the surfaces of iron-oxide minerals. This is released into solution under the highly reducing conditions present in the aquifers and accumulates because of the low hydraulic gradients and small degree of flushing of the aquifers.

Although the potential exposure to arsenic has increased in areas such as southern Bangladesh through increased use of groundwater, relatively few aquifers globally are affected by such serious arsenic problems. Those that are can be reasonably anticipated by an assessment of local geological and hydrogeochemical characteristics supplemented with systematic sampling programmes. It would therefore be unwise to abandon groundwater resources in developing countries in favour of alternatives such as surface water without initial hydrogeological assessment. The potential risks from exposure to inorganic toxins, such as arsenic, as a result of switching from bacterially polluted surface water to groundwater should be weighed against the benefits from the significant reduction in incidence of diarrhoeal diseases.

Source: British Geological Survey/ Mott MacDonald, "Groundwater studies for arsenic contamination in Bangladesh", a British Geological Survey technical report.

shallow aquifers near Sana'a, Yemen's capital, is expected to be depleted within a few years. This contrasts with rising water levels due to sewage infiltration under the city itself (see box 16). With support from the donor community, wells of up to 2 km depth are now being drilled for water supply to Sana'a. Under the North China plain, one of China's most populous and productive regions, groundwater levels are declining by 2 to 3 metres a year in some locations in response to extraction rates estimated at 20,000 million cubic metres a year in the early 1980s (United Nations, 1995).

As mentioned above, the scale and rate of groundwater abstraction are directly related to the massive expansion in pumping capacity that has occurred over the past five decades in many parts of the world. The number of diesel and electrical pumps in India jumped, as previously noted, from 87,000 in 1950 to 12.58 million in 1990 (Central Ground Water Board, 1996) and an estimated 20 million now. In the United States, extraction of fresh groundwater increased from an estimated 47 km^3 a year in 1950 to 120 km^3 a year in 1980. While extraction recently declined to an estimated 110 km^3 a year in 1990, that rate is still more than double the rate in 1950 (Gleick, 1993, table H.12, p. 396).

Intensive abstraction for irrigation and domestic use on the basalt plains south of Asmara, Eritrea

Box 12 Groundwater declines in the Huang-Huai-Hai Plain of northern China

The Huang-Huai-Hai Plain occupies some 400,000 km^2 of north-eastern China, straddling the provinces of Hebei, Henan, Shandong, Jiangsu and Anhui and including the municipalities of Beijing and Tianjin. In 1990, the Plain had an estimated population of 150 million. The aquifers underlying the Plain consist of a thick sequence of downfaulted and partially consolidated Cenozoic sediments which yield significant volumes of groundwater for municipal supply and irrigation. The Plain is estimated to have some 180,000 km^2 of arable land, of which 56 per cent is irrigated. Groundwater development in the Plain started in the mid-1940s with the advent of major irrigation projects. By the early 1980s, it had become apparent that, in many areas of the Plain, groundwater abstraction was exceeding rates of natural recharge and that saline intrusion in coastal zones was increasing. In the vicinity of Beijing, water-level monitoring between 1959 and 1980 indicated a decline of 15 m, with localized cones of depression exceeding this. In Hebei Province, over 7,000 km^2 of the confined aquifers had reduced piezometric heads from between 50 and 70m below ground level and current levels of decline were estimated at 2-3 m per year (United Nations, 1995).

While the consequences of uncontrolled groundwater abstraction in the Plain are manifest in the overall decline of piezometric heads, the requirements to deepen wells, land subsidence and migration of low-quality groundwater, some important technical and economic considerations remain. Among others are:

- The number of abstraction boreholes in the Plain is now estimated to exceed 1 million, making a formal approach to comprehensive regulation among all provinces and water management districts an impossibility.
- The conjunctive use of surface water and groundwater for irrigation in the piedmont and middle Plain needs to be managed if it is to be effective in prolonging the benefit of the Plain's groundwater resources.
- Salinization of soils and the need for drainage have yet to be addressed: waterlogging is assumed to be a function of surface flooding, not inadequate drainage.
- The complex layering of saline and fresh water in the coastal zones of the Plain, together with the development of large coastal cities, poses hydrogeological risks (see cross-section below).

Source: British Geological Survey

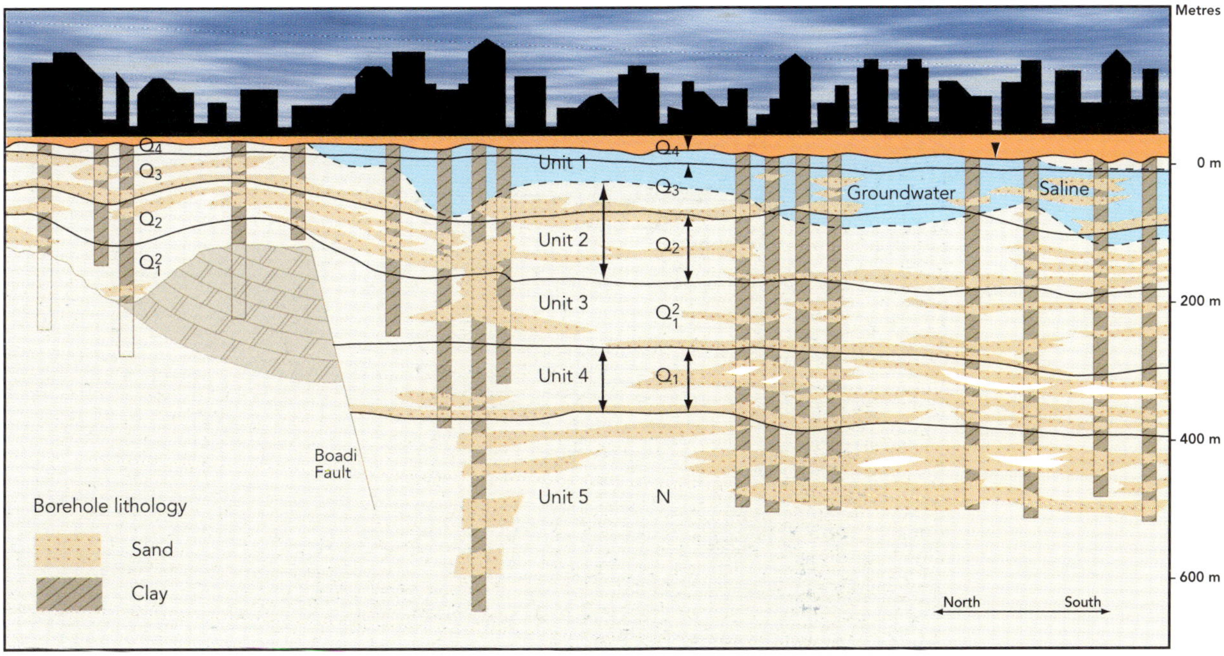

The impacts of long-term abstraction are readily apparent in regions where spring and seepage zones disappear or users have to dig or drill deeper to chase locally falling phreatic or piezometric heads. In addition, aquifer systems themselves are vulnerable to abstraction in many complex, and often not immediately apparent, ways. As in most discussions concerning groundwater over-abstraction, these statistics focus on rates of water level decline and the degree to which estimates of extraction exceed estimates of replenishment, although the provenance of the replenishment, whether recharge from the surface or leakage from adjacent aquifers, is rarely known with any precision. Water level declines can have major social, economic and environmental impacts long before sustainability of the groundwater resource base is threatened in any quantitative sense. The inadequacy of conventional water balance techniques and the need to take into account the actual (i.e., time variant) flow behaviour of the particular aquifer have been highlighted by Miles and Chambet (1995) and are elaborated in box 10.

Discussions of groundwater sustainability need to focus on the ability of the resource to produce key services (including environmental services) plus the economic costs and impacts on equitable access that the loss of those services would entail. Declining water levels, for example, generally have large equity impacts, particularly in the developing world. Wells established for drinking supply often go dry, forcing women and children to walk long distances or wait in line to obtain water to meet domestic needs. Less wealthy farmers are often only able to afford shallow, low-capacity wells. As water levels drop, these farmers can be progressively excluded from access to groundwater by the cost of well deepening and new equipment. This effect undermines the food security and economic development benefits generated by access to groundwater. Where the environment is concerned, water-level declines greatly increase the probability of impacts on streams, wetlands and the occurrence of subsidence. They also increase the

Access to subsidized electricity in Baluchistan, Pakistan, has resulted in a marked expansion of irrigated agriculture and a sharp drop in water-table levels.

Box 13 Groundwater development in the Gangetic basin

The Gangetic basin is filled with unconsolidated alluvial materials to a depth of roughly 6,000 metres and receives an average of 1,500 mm of rainfall each year (Rogers et al., 1989). The total amount of groundwater in storage would be sufficient to meet all needs for hundreds, if not thousands, of years with little danger of physical depletion.

Despite the large stock of water in storage, groundwater development is already having major impacts. In Bangladesh, water-level fluctuations are causing shallow wells to go dry, particularly during summer. This creates major difficulties for villagers in obtaining drinking-water supplies (Sadeque, 1996). It also has major equity effects. Wealthy farmers can afford to deepen existing wells or install new ones; those who are less wealthy often cannot afford the cost of chasing the water-table. Environmental impacts could also be major. As early as 1989, Kahnert and Levine (1989) commented in their summary of a symposium on groundwater irrigation and the rural poor sponsored by the World Bank that "the data show significant seasonal variations in both the water-table and the flow of the Ganges in its lower reaches". Participants in the symposium also expressed concern about the potential impact of increased groundwater extractions on the base flow into the Ganges at low flow periods. Modelling activities currently under way appear to substantiate these concerns. Results suggest that dry-season flows at Farakha Barrage near the Bangladesh border could decline by approximately 75 per cent if historical groundwater development patterns continue (Ilich, 1996).

Declines in dry-season flow are a point of contention between India and Bangladesh. For Bangladesh, these flows are critical for irrigation, for drinking-water supply and for sustaining mangrove areas along the coast. Furthermore, in Bangladesh 70-80 per cent of total animal protein consumption is dependent on fish. As the Eastern Waters Study indicates, activities affecting floods and drainage could interrupt approximately 60 per cent of the nation's fish production (Rogers et al., 1989).

Finally, declining water-tables have major implications for energy consumption. India's base-load electricity deficit now runs at 19 per cent and peak load at over 30 per cent. Much of this is due to agricultural use, primarily pumping for irrigation. In many states, official figures published by the state electricity boards indicate that agricultural demand exceeds 40 per cent of consumption. In some, such as Haryana, it exceeds 50 per cent. While these figures are misleading (they include massive "non-technical" losses), the rate of growth in agricultural electricity consumption has been dramatic. Power for groundwater pumping is highly subsidized. Most commonly, farmers are charged a flat rate based on pump horsepower. As a result, when water-tables fall, farmers have little incentive to reduce extraction. This exacerbates both energy- and overdraught-related problems.

The photo below shows base flows in the upper Ganges basin, Butwal, Nepal.

probability that low-quality water and pollutants will migrate into key freshwater aquifers. Finally, water-level declines can lead to economic exhaustion of the replenishable groundwater resources. As levels decline, drilling and pumping costs increase. Water may still be physically available, but the cost of extraction can be sufficiently high to exclude all but the most high-return applications. The case of the Ganges basin in India and Bangladesh illustrates many of these issues (see box 13).

As the example from the Ganges basin illustrates, *overdraught and water-level declines typically affect the sustainability of uses that are dependent on groundwater long before the resource base is threatened with physical exhaustion.* The sustainability of socio-economic activities in relation to falling groundwater levels is therefore complex. Any analysis of this particular issue needs to look carefully at a range of consequences, including:

- Protection of drinking-water supply sources (both access and quality);
- Equity in access and allocation and poverty alleviation;
- Maintenance of environmental values dependent on groundwater levels or groundwater discharge to watercourses;
- Food security and agricultural production;
- Economic development.

From this consideration it should be clear that the absolute level of extraction in relation to recharge and the rate of water-level change are less important than the uses that are affected by these changes.

Box 14 schematically depicts the gradual emergence of groundwater overdraught and its impact on agricultural economies in India. In early stages, government policies promoted groundwater development as a way of increasing agricultural production and moving the rural economy away from subsistence agriculture. This rapidly acquired a momentum of its own as wealthy farmers realized the benefits irrigation could bring, and the agricultural economy came to depend on groundwater. Uncontrolled expansion then began to have a significant impact on the resource base, and overdraught emerged as a concern. The Government could, however, do relatively little. Policies supporting groundwater development were and remain entrenched and politically popular. Furthermore, the number of wells had increased to the point where regulation of extraction was difficult, if not impossible. During the next phase, as the availability of high-quality

Smallholders growing horticultural produce near urban centres are often the first to lose access to groundwater.

groundwater declines, agricultural growth may decelerate and, in some cases, production may decline. Impacts are likely to be particularly evident in vulnerable regions, such as those subject to saline intrusion or having low levels of recharge and/or storage. The political viability of management may now be increasing, but the human impacts of decreased groundwater availability have already become significant. In some areas, villagers may migrate. Small farmers and other users who are dependent on agricultural labour as a source of income will be particularly hard hit.

4.2 Groundwater extraction and migration of low-quality water

Groundwater flow is slow and governed by two factors: hydraulic gradients and the conductive capacity of the material through which the water is flowing. Water flows down hydraulic gradients from locations of high potential to points of lower potential (points where piezometric levels are respectively higher and lower). Pumping of groundwater may change natural hydraulic gradients and affect both groundwater flow patterns and the natural distribution of discharge and, in some cases, recharge within an aquifer. As a result, even if water levels return to their original elevation when pumping ceases, the migration of lower-quality groundwater or surface water into the aquifer system can occur.

Vulnerability to declines in groundwater quality as a result of increased groundwater extraction is particularly high in certain contexts. These include:

- ***Coastal zones.*** Intrusion of saline ocean water is a common result of pumping, particularly in locations where sediments are highly permeable and in small islands and atolls.

Combining low-head, low-discharge solar pumps with infiltration galleries has allowed skimming of the shallow freshwater lenses in the outer atolls of Kiribati.

- ***Interbedded high- and low-quality aquifers.*** In many locations aquifers containing high- and low-quality water are interlayered. Pumping from high-quality aquifers results in vertical migration of poor-quality water from aquifers which are above or below the point of extraction.
- ***Locations where low-quality water is present on the surface or in adjacent rock formations.*** Pumping often causes lateral migration of low-quality water from adjacent aquifers. As discussed in section 4.1, it also encourages seepage of sewage and return flows from irrigation into aquifers. This often results in microbiological contamination as well as the introduction of nutrients and other contaminants (pesticides, fertilizers etc.).
- ***Locations where rock formations encourage rapid flow.*** Water flows much more rapidly through karstic limestone or other rock formations where large interconnected fractures or cavities are present. These locations tend to be much more vulnerable to rapid contamination from chemical and bacterial sources, for two reasons. First, groundwater flow is relatively rapid, which permits chemical conservation and bacterial survival. Second, because the flow is through relatively large channels, the filtering effect inherent in flow through porous media is reduced or absent.

Karst outflows in Guizhou Province, China

- ***Locations where the geochemistry of adjacent waters and/or the geological formations is incompatible.*** Groundwater geochemistry often differs. This can result in a wide variety of chemical reactions when water containing different levels of key constituents or having differing pH or redox potentials is drawn into and mixes with water in pumped aquifers. Such elements as sulphur, boron, fluoride or arsenic may go into solution, leading to substantial quality deterioration with respect to specific uses.

Figure 6 Groundwater-table decline in the Bangkok metropolitan area

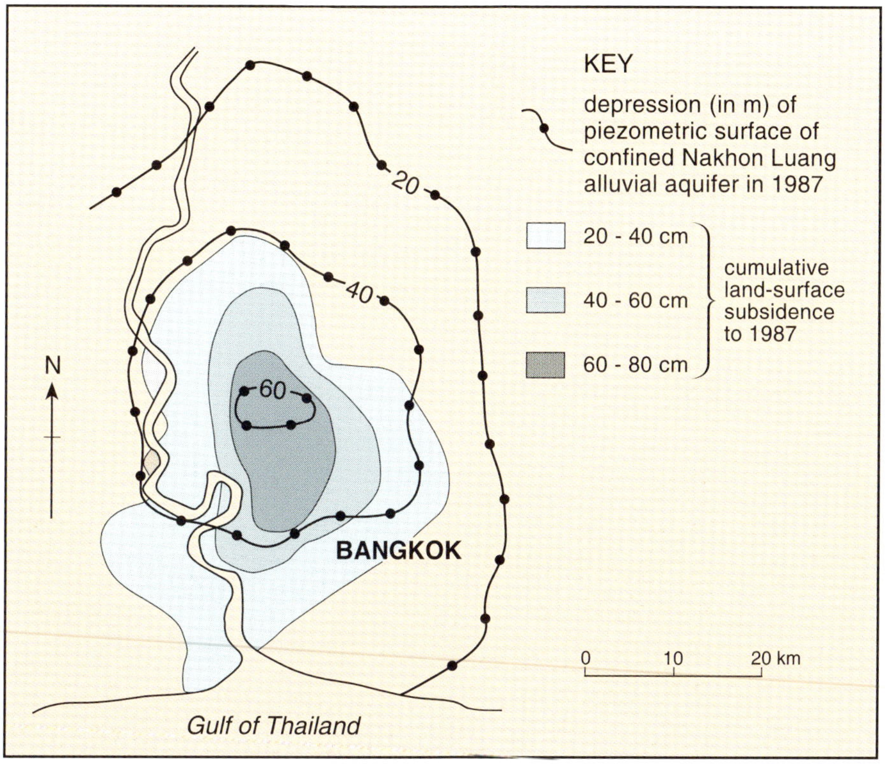

Source: British Geological Survey

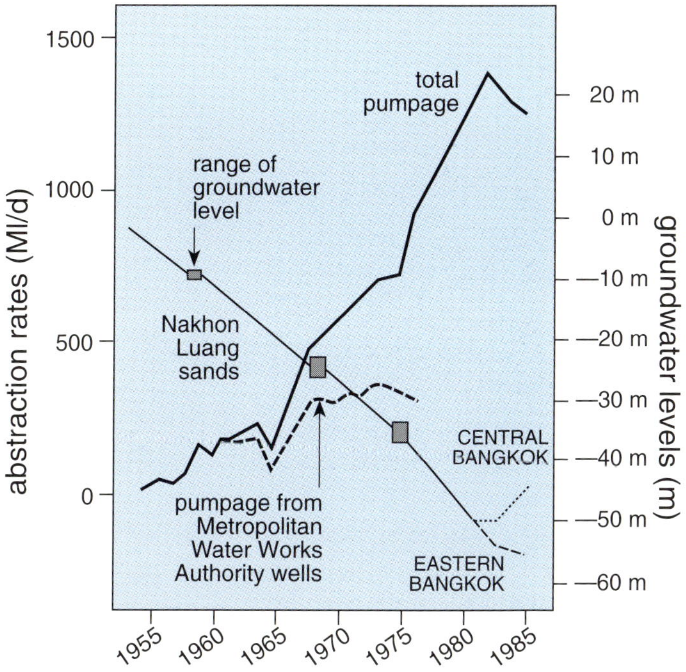

Box 14 Stages in the rise and decline of groundwater socio-economic systems: the Indian experience

	Stage I	Stage II	Stage III	Stage IV
Stages	Popularization of green revolution and tubewell technologies	The rise of the groundwater-based agrarian economy	Early symptoms of groundwater overdraught	Decline of groundwater availability with impacts on local economy and quality of life

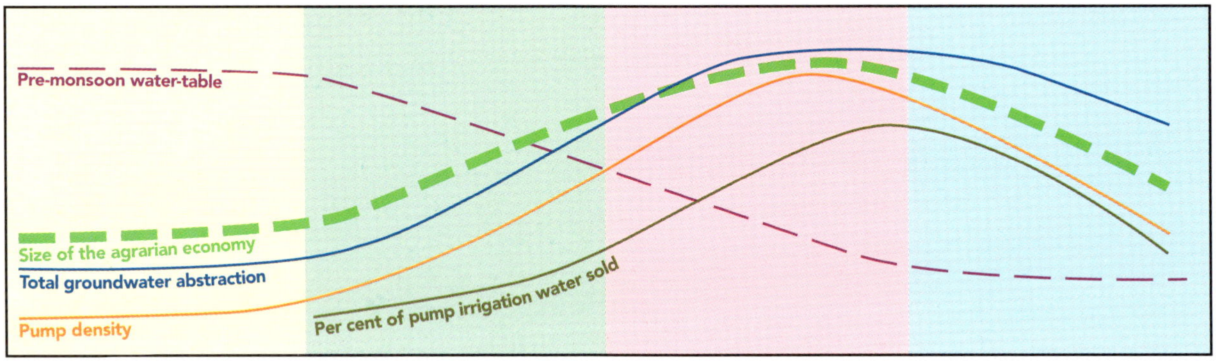

	Stage I	Stage II	Stage III	Stage IV
Example regions	• North Bengal and West Bengal	• Eastern Uttar Pradesh • Western Goadavari • Central Gujarat	• Haryana • Western Uttar Pradesh • Combatore and Madurai, Tamilnadu	• North Gujarat • Coastal Tamil Nadu • Coastal Saurashtra • South-west Bengal
Characteristics	• Subsistence agriculture • Protective irrigation • Traditional crops • Concentrated rural poverty • Traditional water-lifting devices using human and animal power	• Pump ownership skewed • Access to pump irrigation prized • Rise of primitive pump irrigation "exchange" institutions • Decline of traditional water-lifting technologies • Rapid growth in agrarian income and employment	• Diversification of agrarian economy • Long-term decline in water-tables • The groundwater-based "bubble economy" continues booming • Tensions between economy and ecology begin to surface as pumping costs increase and water markets become oppressive • Private and social costs of groundwater use finally part ways	• The "bubble" bursts; agricultural growth decelerates and then declines • Out-migration and poverty increase; in some cases, villages are abandoned • Water-quality problems assume serious proportions • The well-off and the "smart" begin moving out long before the crisis has deepened; the poor are the last to leave and the hardest hit
Public policies and other external interventions	• Targeted subsidy for pumps • Public tubewell programmes • Electricity subsidies and flat tariff	• Subsidies and institutional credit for wells and pumps continue • Donors augment resources for pump capital • NGOs promote small-farmer irrigation as a livelihood programme	• Subsidies, credit, donor and NGO support continue • Licensing, siting norms and zoning system are created but weakly enforced • Groundwater irrigators emerge as powerful vote bank	• Subsidies, credit and donor support go • NGOs and donors emphasize conservation • Zoning restrictions begin to be enforced • Water imports begin for domestic needs • Public- and NGO-sponsored ameliorative action starts

Ion exchange filters at hand-pump headworks in the Malawi rift to remove excessive fluoride

Often several of the above characteristics are present at a single location. Beyond the vulnerability of different regions, however, it is important to recognize that quality deterioration is an unavoidable result of use. In the case of irrigated agriculture, for example, each time water is used, evaporation occurs and salts are concentrated and the return flows result in indirect recharge by progressively lower-quality water.

The impact of quality declines on the sustainability of uses that are dependent on groundwater can be as great as, or greater than, the direct impact of groundwater overdraught. Quality declines can effectively reduce and even destroy the value of entire aquifers as a source of water. Saline intrusion, for example, has virtually eliminated irrigated agriculture and forced the importation of drinking water along parts of the coast of Gujarat, India. The impact on coastal vegetation is equally great. Key environmental resources, such as mangrove wetlands, can be devastated. In inland areas, increases in "natural" contaminants related to pumping have had major health impacts. Increases in fluoride above acceptable levels in drinking water have, for example, been directly correlated with pumping rates, water-level declines and water-level fluctuations in projects in Gujarat, India (Kingdom of the Netherlands, 1992; Wijdemans 1995). Skeletal fluorosis causes joints to "freeze", crippling many individuals in affected areas.

Hence, quality declines associated with increases in groundwater extraction can have as large an impact on the sustainability of key uses as more publicized problems, such as groundwater overdraught.

4.3

Rising water levels and waterlogging

Excessive recharge and poor drainage leading to rising water levels, waterlogging and associated salinization are as major a concern as overdraught in many countries, including India, China, Egypt and Pakistan (FAO, 1997). Rising water levels are, however, common in the command areas of large surface irrigation systems. They are also common in urban areas supplied through imported water supply and where sewer systems have yet to be developed. Finally, it is important to note that groundwater-level rises can be caused by land-use changes—such as forest cutting—with major environmental consequences (see box 17).

Waterlogged area, Haryana State, India

Surface irrigation not only introduces new water but can also obstruct drainage.

4.3.1

Waterlogging induced by irrigation

Pakistan and India contain some of the most extensively documented cases of irrigation-induced waterlogging and salinization (World Bank, 1995; 1998b). Even there, however, it is difficult to evaluate the extent of problems on the basis of available figures. In India, the total area affected by waterlogging due to both groundwater rises and poorly controlled irrigation was estimated in 1990 at 8.5 million hectares by the Ministry of Agriculture (Vaidyanathan, 1994). In contrast, estimates made by the Central Water Commission for 1990, which considered only areas affected by groundwater rises, totalled 1.6 million hectares. Regardless of the actual extent, waterlogging problems represent a major surface-water and groundwater management challenge—and one that cannot be addressed in the absence of an integrated approach that incorporates surface-water imports and use as well as groundwater. Large areas in Pakistan face similar problems (see box 15).

Box 15 Groundwater management in the Indus basin

Recent estimates are that irrigated land furnishes 90 per cent (by value) of Pakistan's agricultural production, accounting for 26 per cent of GDP and employing 54 per cent of the labour force. It is apparent on the evidence of current investments that priorities in irrigated agriculture easily outweigh other interventions. Maintaining a bank of soil resources and a flow of water resources to support food production for a population growing at some 3.0 per cent a year has become an imperative for Pakistan. The bulk of this productivity is associated with the Indus basin.

The Indus basin is filled with thick alluvial sediments deposited by the Indus River and its five main tributaries, Jhelum, Chenab, Ravi, Sutlej and Beas, forming a thick set (300 to 500 metres) of unconfined and leaky aquifers. Before the introduction of a weir-controlled canal irrigation system, the groundwater-table was relatively deep under most of the plain. As a result of the additional recharge introduced by irrigation, the water-table started rising at a rate of 15 to 75 cm per year. The position of the water-table before and after the introduction of the large canal networks in the upper part of the basin rose 20 to 30 m in 80 to 100 years. The quality of groundwater varies in vertical and horizontal directions and is related to recharge of the aquifer; in general, water from shallow wells located near sources of recharge is of good quality. Along the rivers and in the upper reaches of the doabs where precipitation is a major source of recharge and maximum canal supply is available, groundwater usually contains less than 1,000 ppm (parts per million) of dissolved solids.

The Indus basin was developed through surface irrigation under British colonial rule in the late nineteenth century, but the threat to the system of saline accumulation in irrigated soils was appreciated by the original design engineers. What were the results?

- Public tubewell development started in the 1960s through Salinity Control and Reclamation Projects (SCARPs). Since drainage projects alone generally have a low economic rate of return, priority has been given to locating SCARPs in areas of usable-quality groundwater. As a result, 90 per cent of the SCARP tubewells and 95 per cent of pumped groundwater is from freshwater groundwater zones. In effect, SCARPs have evolved into groundwater-supply projects in which drainage is a by-product.
- Capacity of the private sector to develop good-quality groundwater (which was not appreciated in the early planning stages) was triggered by the SCARP development.
- The salt balances of the Indus and its associated sub-basins have been disrupted as the hydrochemical systems have progressively become closed and the supplemental generation of salt through waterlogging has further exacerbated the positive salt balance.
- The Indus is effectively a saline sink with minimal flushing and outflow. This applies to the Indus plain as much as to the North-west Frontier Province and Baluchistan sub-basins, which are also in danger of becoming closed subsystems.
- The recently launched National Drainage Programme has dropped subsidies from public tubewells in fresh groundwater areas.

The physical and chemical environment in which groundwater is found and is evolving is complex, particularly in the shallow layers that have experienced recent groundwater recovery and quality changes. Relatively fresh groundwater occurs side by side with saline groundwater or under or overlain with saline groundwater. This requires a high degree of operational knowledge in the management of groundwater in order to ensure its sustainability in terms of quantity and quality. Identification of hydrogeological processes and the establishment of a physiographic framework is therefore imperative to both explain and quantify the groundwater occurrences and the rate of aquifer replenishment and depletion.

Rising water levels in the command of surface irrigation systems have fundamental implications for the sustainability of social objectives that are groundwater-dependent. In the case of food security, estimates indicate that irrigation-induced salinity and waterlogging reduce crop yields in Pakistan and Egypt by 30 per cent (FAO, 1997; World Bank, 1994). In India, the problem is serious enough to threaten growth of the agricultural economy (Joshi et al., 1995). The impact of waterlogging and salinization on farmers and regional economies can be insidious. In the initial years, the introduction of irrigation often causes a dynamic transformation of regional and household economies. Farmers shift to high-yielding varieties of grain and are able to grow valuable market crops. Wealth is created. As the water-table rises, however, this "bubble economy" based on unsustainable water management practices slowly deflates. Land and the unsaturated zone of the soil, once salinized, are difficult and expensive to reclaim. Ultimately, many farm families—and regional economies—may be worse off than before the introduction of irrigation unless sustainable and affordable methods of remediation can be found.

4.3.2 Water-level rises under urban areas

Water-level rises are a major feature in many urban areas, particularly once cities begin to rely on imported supplies. Although urbanization may reduce direct infiltration of rainfall because of large impermeable areas that are created, recharge below cities is often far higher than pre-urban levels (Morris et al., 1994; World Bank, 1998b). In a recent study, the increases in recharge under Mérida, Hat Yai and Santa Cruz (cities in Mexico, Thailand and Bolivia, respectively) ranged from 130 per cent to 600 per cent. In Lima, Peru, recharge has increased from essentially zero to 700 mm a year (Morris et al., 1994). Since most of this recharge comes from leaking sewers and water mains, the potential for pollution is high. Such implications are discussed in detail in the next section.

Whether increases in recharge under urban areas lead to rising groundwater levels depends heavily on extraction rates and local hydrologic conditions. In many cases, water levels fall during the initial phases of urbanization when cities depend essentially on groundwater sources for water supply. As locally available groundwater resources are depleted or become polluted, urban areas often import water supplies. This can lead to a reversal of declines and substantial rises in groundwater levels.

When water imports induce rising water levels in unconfined aquifers, the effect enables shallow wells to serve as a major source of water supply for the poor. Since, however, pollution levels are generally much higher in shallow urban aquifers, particularly in areas not served by sewer systems, those dependent on shallow wells face major health risks. This is well illustrated by the case of Sana'a in Yemen where, despite general conditions of overdraught in aquifers supplying the city, water levels under the city itself are rising (see box 16). Furthermore, high water levels under urban areas cause drainage problems, leading to the creation of stagnant and highly polluted surface-water bodies. These become centres for downward contamination and sources of disease. Health and economic impacts tend to be concentrated in poor urban areas, where residents frequently lack the political pull or financial resources to obtain access to piped water supply and sewage systems.

Box 16 Protect surface water now or groundwater later? The quandary in cities with poor sanitation: Sana'a, Yemen

Sana'a, capital of the Yemen Republic, lies in the centre of an intermontane basin, over 2,200 m above sea level, surrounded by spectacular volcanic peaks. The city is underlain by alluvium, resting on a Cretaceous sandstone which forms the main aquifer. With the architectural treasures of the old city (a UNESCO world heritage site) at its centre, Sana'a is growing rapidly, with a population increase of over 10 per cent a year and a rising demand for water. A detailed study (Alderwish and Dottridge, 1998) shows that urban recharge forms the main component of total recharge.

Public water supply is drawn from two well fields in the sandstone outside the city and serves only 30 per cent of the 972,000 inhabitants (1994 census). Their water use is estimated at 100-120 litres a day per person, with 30 per cent distribution losses. The remaining 70 per cent obtain water by pipes or tanker from private suppliers whose boreholes are located in the urban area. The water use of this group is lower, 60-80 litres a day per person, and because of short pipe lengths, losses are lower, typically 20 per cent. The city is largely unsewered, with only 12 per cent of households connected to the sewerage system. This water undergoes only very basic treatment at a plant outside the city before being discharged as surface flow and posing a significant health risk to the local villagers. Discharge from cess pits forms the main component of urban recharge, contributing 12.5 million m^3 in 1993, which comprises almost three quarters of the estimated urban recharge. Clean water from mains leakage provided 3.4 million m^3, with 1 million m^3 from industry and local irrigation, and additional recharge from infiltration during sporadic floods in the main wadi crossing the city.

As a result of urban recharge, groundwater levels in the city are stable or rising, especially in shallow wells, in contrast to the basin trend of declining levels. The alluvial aquifer shows clear signs of contamination, with elevated nitrate and chloride beneath the more densely populated areas. Similar trends are observed along flow lines in the underlying sandstone aquifer, although no bacteriological contamination has been detected in deeper boreholes. Both chemical trends and piezometric levels indicate downward leakage from alluvium to sandstone, which allows transfer of recharge to the main aquifer, but also provides a pathway for pollutants. Although this poses a potential problem for the future, at present cess-pit disposal causes minimal immediate danger to public health because of the current lack of adequate sewage treatment facilities.

Water supply and disposal issues/conflicts

- Fast-growing city, semi-arid location, poor water infrastructure
- More than 80 per cent of the population uses on-site sanitation; this maintains water levels but is adversely affecting water quality and prejudicing resource sustainability
- The regional tendency towards over-exploitation limits the scope for extension of the currently low public supply coverage (30 per cent) to substitute for local supplies as quality declines
- Extension of sewerage without commensurate investment in sewage treatment in the intermontane basin would seriously aggravate the existing health risk to the downstream population

Some water management options

- Extension of sewerage and significant improvement of sewage treatment: trade-off of improved recharge quality against risk of water levels falling for private suppliers too
- Investigation of the potential for shallow groundwater use in the city, e.g., for recreational-area irrigation
- Increased scope for recharge of higher-quality water outside city limits to provide lateral dilution, e.g., wadi spate-detention basins to induce recharge
- Minimal scope for demand management, as per capita supply to both the publicly and privately supplied consumer is already low

4.3.3 Water-level changes in response to vegetation cover

Beyond the issue of water-level rises in the context of surface irrigation systems or urban water-supply imports, land-use changes can have a significant impact on groundwater levels. Forest and vegetation cover has long been recognized as a major factor influencing run-off, infiltration and evapotranspiration from shallow water-tables. Watershed treatment, involving the establishment of tree, bush and other plant cover, is widely used as a way to reduce run-off and increase infiltration. This is frequently assumed to increase recharge. Documents prepared by the Integrated Watershed Development Program in Rajasthan, for example, specifically mention increases in groundwater recharge as a result of watershed treatment. As box 17 documents, however, the effect of surface vegetation on groundwater levels is not automatic and depends on the balance between improvements in infiltration caused by increased vegetation and relative changes induced in evapotranspiration. In some cases, removal of forest cover has caused water levels to rise significantly, with major environmental consequences.

The result of dryland salinity in New South Wales, Australia

Box 17 Vegetative impacts on water-table depth: examples from Australia

Under natural conditions, the depth of the water-table below ground surface is determined by the long-term balance between aquifer recharge and discharge. In many countries, increased discharge (abstraction) has led to falling water-tables, but changes to recharge can also be a problem. Recharge is determined mostly by the balance between rainfall, evaporation, evapotranspiration, and run-off in the soil zone. Changing the natural vegetation cover will usually lead to a change in this balance.

In Australia, large areas of natural forest and scrub have been cleared since European settlement and replaced with grass or crops. This has resulted in a significant increase in aquifer recharge, causing water-tables to rise to a new equilibrium depth at a shallower level. The response to land clearing typically occurs within a few years, with farmers noting increased yields from wells and even the development of artesian conditions in some valley floor bores. Local streams also develop higher base flows.

An unforeseen side effect of the new equilibrium has been the saturation of salt-bearing clays in many valleys and an increased flux of groundwater through these clays as the groundwater system responds to the increased recharge. The increased flux of groundwater leads to destabilization of the clays and to the accumulation of salts in the shallow soil. Vegetation becomes stressed by the salt accumulation and dies. The result is rapid sheet and gully erosion over areas associated with groundwater discharge. The erosion products have led to increased turbidity and salt loads in the streams which drain these areas.

The impact of the development of dryland salinity in the headwaters of many streams in the Murray-Darling basin is a major concern of the Murray-Darling Basin Authority, which has the task of managing water quality and quantity for the extensive irrigation areas lower down the catchment.

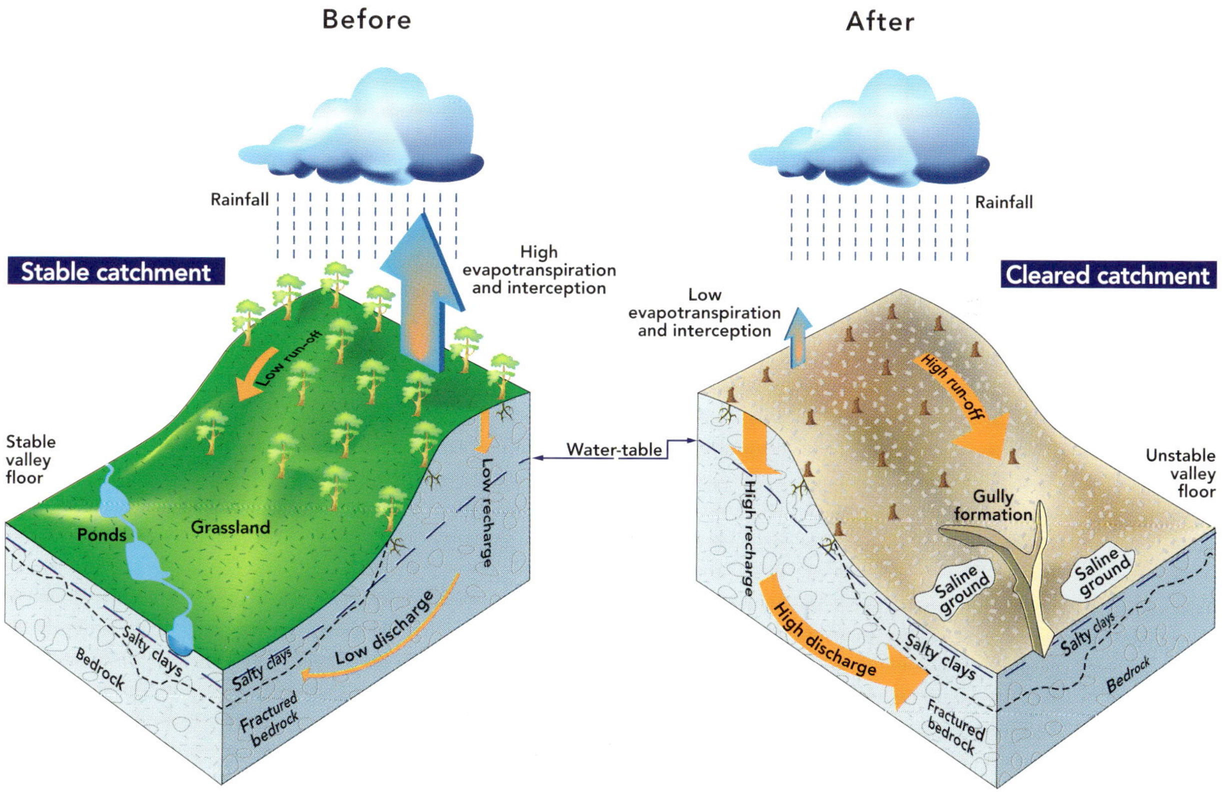

4.4 Pollution

Pollution is widely recognized as one of the most serious challenges to sustainable management of groundwater resources. The significance of pollution for groundwater resources is greatly increased by the long timescale at which processes affecting groundwater function. As Morris et al. (1994) comment: "It is important to appreciate the differences between surface-water and groundwater systems. In the former, the water is typically being replenished, at least in the case of rivers, within times-cales of weeks or at most months. Replenishment times for groundwater systems are very much longer. This is because water usually takes many years to move through the soil and unsaturated zone of the aquifer. Once there, it can take a further period of many tens or hundreds of years to flow into a supply borehole." In some of the deeper aquifers, groundwater is likely to be thousands of years old (Edmunds and Wright, 1979; Edmunds et al, 1987). In addition to the relatively slow movement of water in many aquifers, rocks and soil absorb and otherwise attenuate the presence of pollutants. Not all aquifers are equally vulnerable to pollution. Those where fractures or cavities permit rapid flow tend to be more vulnerable than those where water flows slowly through porous media and more opportunities exist for attenuation of pollutants. Vulnerability to pollution has, however, an inverse relationship to the difficulty of remediation. Once goundwater is polluted, its slow movement through a porous aquifer generally makes clean-up difficult, expensive, and in some cases impossible. Groundwater pollu-tion can, as a result, cause damage to groundwater resources that is virtually irreversible. The relative vulnerability of aquifers to pollution is illustrated in box 18.

Beyond the inherent vulnerability of aquifers to contamination, much depends on the nature of pollutant sources. Contaminant behaviour varies greatly with respect to the specific transport properties in each aquifer system. In addition, the range of contaminant types is increasing as new products appear in effluent disposal and land application. Three main sources of groundwater pollution are discussed here: agricultural, urban and industrial.

4.4.1 Agricultural pollution

Although groundwater pollution from agricultural activities is a growing concern worldwide, its extent is very poorly documented. In the United States, over 44 states cite agriculture as a cause of groundwater pollution, listing in aggregate sites covering 147 million hectares as affected (Gleick, 1993, p. 256). The widespread nature of groundwater pollution related to agriculture in the United States is linked to the relatively long history of chemical fertilizer and pesticide use. In many developing countries, agricultural chemical use has, at least until recently, remained low in comparison to levels in industrialized countries. This may no longer be the case, particularly in such countries as India and China, where irrigation is extensive. Concerns over groundwater pollution from agricultural chemicals were raised as a major issue in India a little over two decades ago (Chaturvedi, 1976) but few data were available. At that time, the level of agricultural chemical use was very low. By 1991, however, fertilizer use per hectare of agricultural land was 60 per cent *higher* than in the United States (Repetto, 1994, emphasis in original). At present, no agency in India has a systematic programme for monitoring potential non-point sources of pollution. Fragmentary data indicating the poten-

Box 18 Aquifer vulnerability and pollution risk

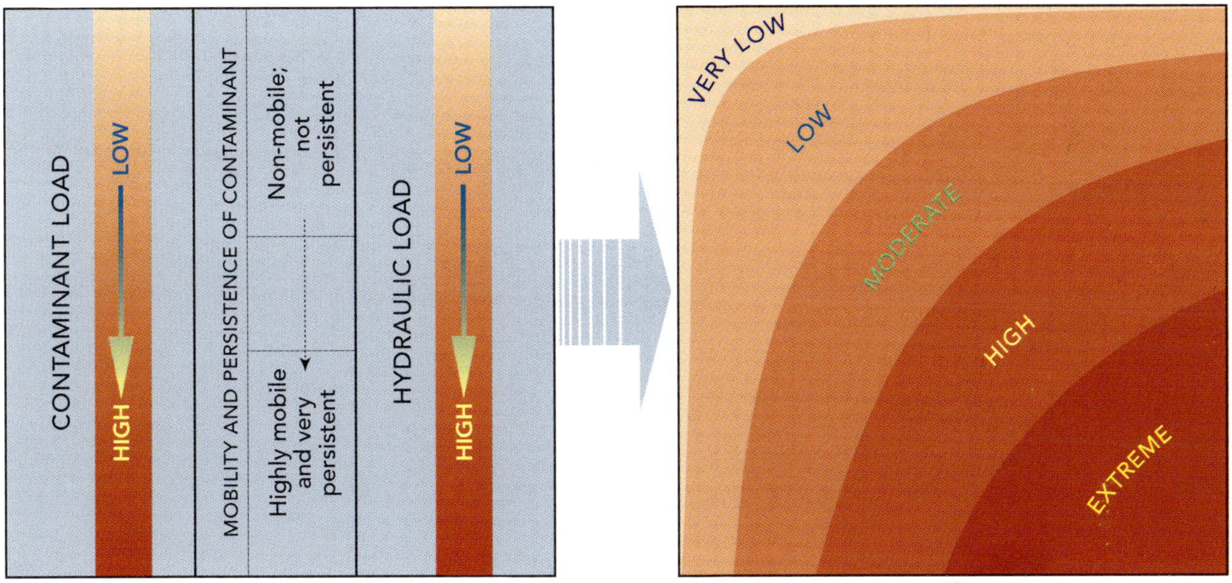

Groundwater pollution risk can be conceived of as the interaction between two semi-independent factors:

- The contaminant load that is applied to the subsurface as a result of human activity;
- The natural vulnerability of the aquifer to pollution.

Adopting this scheme, it is possible to have high vulnerability but no pollution risk (because of the absence of significant load) and vice versa. Both are perfectly consistent in practice. It is important to recognize that the contaminant load can be controlled or modified but not aquifer vulnerability.

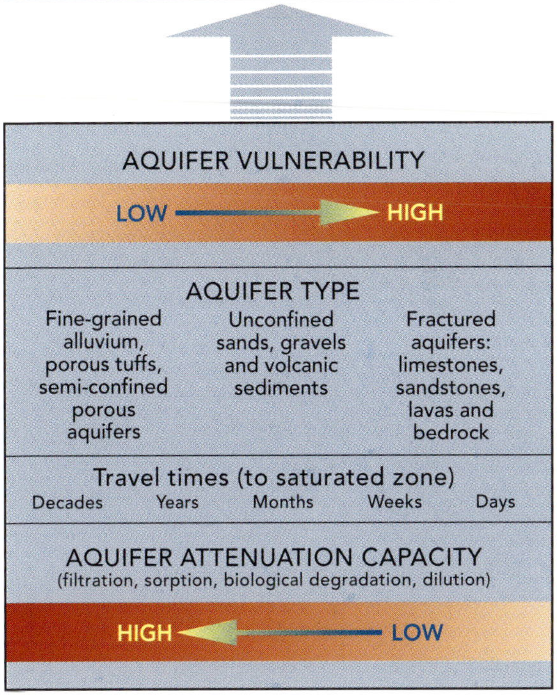

Source: British Geological Survey

tial extent of agricultural pollution problems are, however, available. In Gujarat, for example, maps prepared by the Central Ground Water Board (CGWB) show nitrate concentrations exceeding 45 mg/l (the maximum for drinking recommended by the World Health Organization) in over 370 sample sites scattered across the state (Phadtare, 1988). How much of this is related strictly to agricultural pollution and how much to domestic or other sources is unknown.

Aside from non-point-source considerations, it is important to recognize that nitrate and other nutrient pollution in groundwater is often related to agricultural practices other than the use of chemical fertilizers. Any location where animal wastes are concentrated, such as feed lots or poultry farms, can release high levels of nutrients into groundwater. In Petaluma, California (United States), for example, poultry waste created plumes of nitrates in the soil and underlying groundwater. Even after poultry operations were stopped, concentrations of over 300 milligrams of nitrate per litre, far exceeding the 45 mg/l maximum for drinking water, were measured in Petaluma's groundwater (California Department of Water Resources, 1994b, p. 129). Nitrates often cause methoglobanaemia, a disease that can be fatal for infants. Nitrate removal is so expensive that affected areas in the United States are often forced to rely on imported or bottled supplies for drinking and food preparation (Department of Water Resources, 1994a).

In addition to fertilizers, pesticides and herbicides are other major sources of groundwater pollution related to agriculture. In some circumstances, soils can absorb or immobilize a large fraction of such agricultural chemicals. Many pesticides and herbicides, however, break down slowly under aquifer conditions and, as a result, can persist over long time periods. In any case, groundwater pollution data are generally scarce, and chemical analysis of water samples needs to be specific to detect their presence.

The dispersed nature of sources of pollutants is a core challenge facing both monitoring and control of groundwater pollution related to agriculture. Unlike industry or municipal sewage systems, agricultural pollutants are dispersed over large land areas. While return flows in drainage canals can be monitored, it is difficult to determine the extent of direct seepage of pollutants through soils and into the groundwater until contaminant concentrations in groundwater become significant.

4.4.2 Urban groundwater pollution

As previously noted, groundwater recharge in urban areas is generally much higher than under natural conditions. The additional recharge is derived principally from leaking sewers and other waste-water sources. Broken sewers in the United States are estimated to lose 950 million cubic metres of waste water each year (Pedley and Howard, 1997). Much, if not most, of this represents polluted recharge to groundwater. Direct leakage of waste water to groundwater in developing countries is probably much higher. In many cities, a large portion of the waste water generated is discharged directly into unlined canals. Where sewer systems exist, leakage levels are almost certainly much higher than United States rates due to lack of resources for maintenance, variability in construction materials and absence of adequate treatment facilities. Furthermore, in many urban and peri-urban areas, pit latrines and soak pits are used to dispose of domestic waste water. These are often relatively deep (greater than 3 metres) and discharge wastes below the soil and weathered zone layers that have the greatest capacity to filter, absorb and otherwise attenuate pollutant concentrations (Pedley and Howard, 1997).

Box 19 Urban pollution: the Hat Yai and Santa Cruz cases

The cities of Santa Cruz in Bolivia and Hat Yai in Thailand are largely unsewered. As a result, significant quantities of domestic and some industrial wastes are discharged to the subsurface. Principal contaminants entering the subsurface in this way include nitrogen, chloride, long-chain organics and faecal coliforms, and shallow groundwater beneath both cities is highly polluted with these contaminants. Metals are generally absent, as the prevailing pH of the groundwater in both cities is outside the range of predicted metal mobility. Synthetic organics are also largely absent in the waste water and groundwater.

Both cities depend on groundwater from deeper semi-confined aquifers, and pumping from these aquifers has induced downward leakage. Elevated levels of nitrogen and chlorine in some of the deeper groundwater indicates the depth to which modern recharge has penetrated. Faecal coliforms and elevated DOC (dissolved organic carbon) are generally not observed in these deeper groundwaters, indicating attenuation at shallower depths. In the case of Hat Yai, the most polluted urban groundwaters have chloride concentrations indicating that they are largely derived from canal seepage and occur where groundwater abstraction and, consequently, downward leakage are greatest.

At present, water from deeper aquifers remains excellent. Continued high levels of extraction will, however, cause further downward leakage of waters with high levels of nitrogen and chlorine from upper polluted levels. Individuals and communities depending on shallow wells for domestic water supplies face significant health risks from pathogens as well as other pollutants. Although pathogens are attenuated at upper levels, downward leakage of water containing elevated levels of nitrogen and chlorine will ultimately increase health risks for the entire population dependent on water supplies from aquifers underlying both cities.

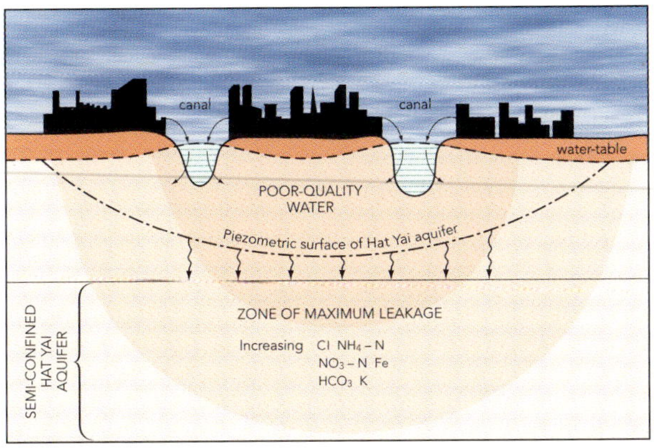

Mixing of unpolluted regional groundwater flow and canal seepage in Hat Yai, Thailand

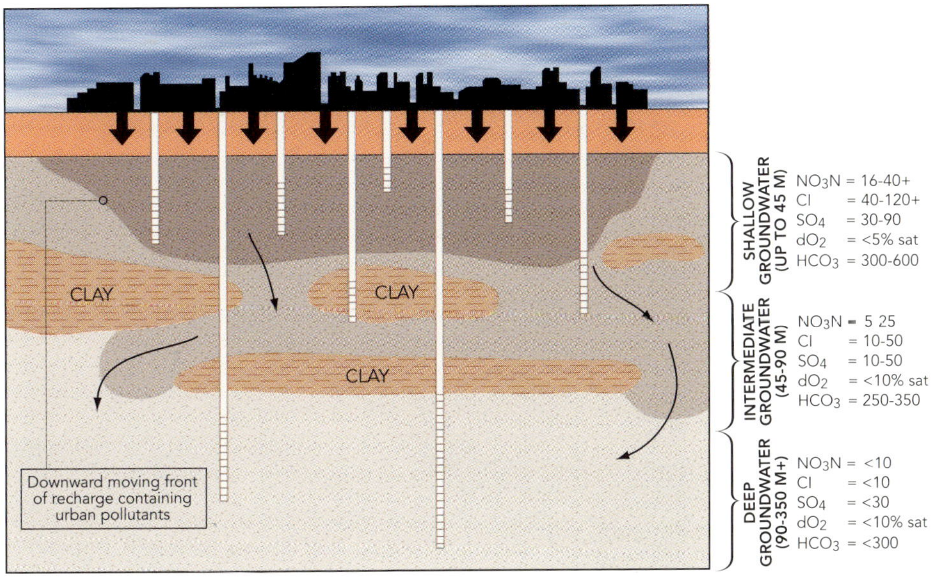

Schematic cross-section of Santa Cruz, illustrating main groundwater-quality variations

Source: British Geological Survey

Unlined sewage and storm flow drain in Sana'a, capital of Yemen

The impact of urban waste-water discharges on groundwater is well illustrated by the cases of Santa Cruz, Bolivia, and Hat Yai, Thailand (see box 19). In both these cities, direct discharge of untreated waste water has led to substantial increases in pollutants (NO_3, NH_4, Cl, faecal coliforms, and dissolved organic carbon) in the shallow aquifers. The quality of deeper groundwater is still good, but pollution fronts are moving downward in response to extraction from deeper levels for drinking-water supply and other uses (Morris et al., 1994). This situation is typical of many cities, particularly in rapidly urbanizing sections of the developing world.

The potential impact of waste discharge on chemical contamination of groundwater by nitrates and other compounds tends to be recognized by water-supply officials. It is often assumed, however, that the filtering action of aquifers and relatively long residence times underground are sufficient to remove pathogens except where open or poorly sealed wells are directly contaminated by surface-water inflows. This perception is inaccurate. According to Pedley and Howard (1997, p. 182): "Bacteria can survive up to 50 days or more in sub-surface environments and viruses for far longer. "

Overall, pollution of shallow aquifers under cities represents a major threat to the sustainability of drinking-water supplies in many urban areas throughout the world (British Geological Survey, 1995; World Bank, 1998b). This threat is particularly high where regional hydrogeological conditions permit rapid flow of contaminated water into aquifers and the wells tapping them. Aquifers in karstic carbonate rocks or fracture zones are, for example, far more susceptible to contamination than aquifers where groundwater flows through porous media such as soil or sandstone. The threat is also particularly high where large portions of the urban population dispose of untreated wastes directly through soakaways and latrine pits *and* depend on shallow wells for drinking-water supply.

4.4.3 Industrial pollutants

Pollutants associated with industry and the modern technological society—such as transportation systems based on diesel or gasoline and heavy metals from industries such as textile and paper mills—are major sources of groundwater contamination (Lerner and Walton, 1998; Mather, 1998). Data presented by Gleick (1993) indicate that in the United States between 2.4 million and 4.8 million sites in 39 states have been reported where benzene, toluene, xylene and petroleum products are seeping into groundwater from underground storage tanks. From 25 to 30 per cent of all underground storage tanks may leak. In addition, 29 states report an aggregate of between 10,000 and 16,000 chemical spills each year. Operating and abandoned wells for oil and gas production add another 1.7 million potential sites where brines and petroleum may contaminate shallow and intermediate aquifer systems (Gleick, 1993, table D.15). Leakage from underground tanks is one of the most important sources of pollution. These tanks are often widely dispersed. The California Department of Water Resources, for example, notes that solvent pollution in the San Gabriel Valley "is so widespread in the groundwater that it is generally not possible to identify individual sources" (Department of Water Resources, 1994b, p. 128).

Public attention with regard to groundwater pollution often focuses on "hot spots" where industrial activities have polluted large areas. Sites of this type often receive national attention. Jetpur, a textile town in Gujarat, India, where over

1,200 small industrial units drain effluents containing cadmium, zinc, mercury, chromium and other pollutants into small rivers and thence into groundwater, is a prime example (Moench and Matzger, 1994). Governmental monitoring and clean-up activities also tend to focus on high-profile sites. The "Superfund" sites in the United States and activities of the state and central pollution control boards in India are typical of many governmental initiatives, particularly during early phases, when the significance of groundwater pollution is only beginning to be recognized. In India, the Central Pollution Control Board has a programme to monitor groundwater quality in 22 critically polluted sites (Moench, 1996). There is, however, no baseline monitoring of potential industrial pollutants except within these hot spots.

The "hot spot" focus of public attention and many government initiatives tends to play down the importance of dispersed sources of industrial pollutants, such as trace metals and organic solvents. Many such pollutants have extremely long residence times in aquifers because of their low solubility. Because they do not dissolve rapidly, they can remain indefinitely as a concentrated source of pollution within an aquifer. In some cases, gradual volatilization of organic solvents in aquifers can even become an air-quality hazard. Dispersed sources of industrial pollutants are much harder to identify, monitor and control than the effluent from specific factories or industrial areas. As such, these dispersed sources may well represent a greater threat to groundwater resources than concentrated industrial effluent flows.

Data on groundwater pollution in developing countries, particularly pollution related to diverse point sources, such as mining activities, underground storage tanks and direct discharge of effluent to water bodies and watercourses, are generally unavailable. With increases in transportation needs and industrial activities, the number of sites where pollution is occurring is, however, increasing rapidly.

Pollution from ore processing into karstic base flows in Hunan Province, China

Cattle drinking municipal waste that is being used for irrigation in Ajmer, Rajasthan, India

4.4.4

Implications of groundwater pollution

The full impacts of groundwater pollution on health, agriculture and the environment have never been comprehensively assessed. In the case of health impacts, Pedley and Howard (1997, pp. 180-181) observe that: "The contribution made by contaminated groundwater to the global incidence of waterborne disease cannot be assessed easily; for many countries the incidence of waterborne disease is not known accurately and the data for groundwater usage are not available. Where public health statistics are available, the data are insufficient to determine the source of the water involved in the transmission of the disease." In comparison to other topics, such as the environment, collection of public health data is widespread and, while not without problems, relatively well established. If it is difficult to assess the impact of groundwater changes on health (where at least some data are available), then the magnitude of the challenge in assessing impacts on other values should be clear.

Lack of information on the health and other costs associated with groundwater pollution and quality declines may lead to questions regarding the importance of these problems. While the dangers of pathogenic organisms are widely recognized, in developing-country situations, officials often informally emphasize the lack of evidence that such diseases as methoglobanaemia, or responses to toxic substances, are occurring in any but the most polluted areas.[9] On the basis of this perception, they often advocate relaxation of standards. Part of this response may be due to the widespread incidence of many other health and disease prob-

[9] Interviews with individuals in India, Tunisia and Yemen between 1991 and 1997.

lems, making diagnosis difficult. Part may also be due to priorities. Pollution control and remediation of aquifers are expensive. In the United States, substantial debate has emerged over the cost of clean-up in relation to the value of groundwater resources (National Research Council, 1997). In developing countries, demands on limited financial resources are often more intensive. As a result, more questions arise regarding large investments in pollution control or aquifer remediation that have few immediately observable returns.

The above observations do not imply that the human and environmental burdens associated with groundwater pollution are minor. Diseases related to water pollution are a major concern in many parts of the world. In 1994, cholera caused over 10,000 deaths; during recent years typhoid caused 25,000 deaths, amoebae 110,000, and diarrhoeal diseases 3.2 million among children under five years of age (Pedley and Howard, 1997, p. 181). Comprehensive data on deaths and disease caused by absorption of trace metals and other pollutants are not available. Overall, however, days lost to disease and the continuing burden of sickness on society far outweigh the actual number of deaths. Although the amount of death and disease that can be attributed to groundwater pollution per se as opposed to surface-water pollution is unknown, it clearly adds a continuing burden to the health of large populations, particularly in developing countries. In a similar, but mostly undocumented, manner, groundwater pollution clearly affects a wide range of other key environmental and social values.

5 Root causes of competition and issues for groundwater management

The growing number of wells, uncontrolled pumping and unregulated disposal of pollutants are all proximate causes of emerging groundwater problems. Underlying these immediate causes, however, are more fundamental factors driving gradual degradation of the resource base. This chapter focuses on some of the root causes of emerging groundwater problems. Groundwater is not, after all, an inherently scarce resource. Emerging problems relate more to "management" issues—use efficiency, allocation and understanding—than they do to the ultimate sustainability or carrying capacity of the resource base. The goal of this chapter is to identify the core challenges that management principles have to address.

Population growth, economic expansion, the spread of drilling techniques, pumping technologies and competition represent fundamental changes in the context of groundwater resource exploitation. Society's inability to manage groundwater in ways that reflect these changes in context lies at the heart of current problems. As the context within which groundwater is used changes, a range of factors complicate society's ability to manage groundwater resources in a sustainable fashion. Key factors include: (1) the common-property/open-access nature of the resource base; (2) policy failures and the impact of policies encouraging development; (3) the way society perceives groundwater; (4) the lack of public awareness regarding groundwater; (5) great variability in groundwater conditions, problems and social contexts; (6) the limited capacity of governments and other organizations to control groundwater use and pollution; and (7) the absence of data on the resource itself and scientific understanding of the processes that govern and condition groundwater flow and availability.

5.1 The common-property nature of groundwater

Throughout the West, legal traditions stemming from Roman common law often underlie a direct link between groundwater rights and landownership. This is also true in many parts of the world where legal traditions stem from other roots. This point is elaborated in section 6.1, but it is sufficient here to observe that under common-law systems, landowners have a right to dig or drill wells and exploit groundwater for use on their own lands. As Barraqué (1997b) notes, this traditional view of groundwater resources was "linked to the belief that they were plentiful wherever they could be found, and that they could be treated as *res nullius.* But this is changing now: groundwater is increasingly considered as the rest of water resources, i.e., a *res comunis omnium.* This makes public regulation and management easier to develop, even though the lack of a full recognition of the notion of common property in positive law makes the picture quite complex." While this use of terms reflects a sociological interpretation rather than its legal condition (*res nullius* would apply to wandering beasts, for instance, and groundwater would be more strictly defined as *res privae*), it accurately reflects the evolution of social thinking through time and it highlights the current public nature of groundwater, i.e., groundwater as *res publicae.* Of particular concern is that common customary uses in many developed and developing countries are not protected by clear administrative rights. Yet a variety of means, including public domain, police power, and expanding doctrines such as the concept of public trust, provide suitable legal foundations for the protection of groundwater resources and also of legitimate and beneficial uses. Still, it has to be emphasized that many countries, particularly developing ones, lack the means to articulate the protection of common customary uses of either groundwater or groundwater-fed sources. This is both a legal and a political issue, since this lack of means includes a lack of substantive legislation for the protection of common customary interests, a lack of standing to act, and a lack of administrative and judicial forums to which to resort. As a result, many local indigenous communities in Asia, Latin America, the Middle East and elsewhere see their fishing, watering, drinking and irrigation uses depleted by pumping for municipal consumption, mining or large-scale farming.

Groundwater resources have inherent common-property characteristics. In most countries, access is limited primarily by landownership and by the landowner's ability to raise the financial resources necessary to drill or dig wells. Aquifers, however, generally extend over more than one land title and are tapped by numerous landowners. Groundwater conditions are consequently heavily influenced by the cumulative actions of sets of individual users and their individual abstraction regimes. Individual landowners have virtually no capacity to influence access to (or use of) the resource base by other landowners. Furthermore, since they have no ability to protect groundwater resources, there is little incentive for individuals to use groundwater efficiently or invest in maintenance of the resource base. Any improvements in resource condition or reserves created by individual users will tend to be captured by other users. In many ways, this represents a classic "tragedy of the commons" as described by Garret Hardin (1968).

In addition to the common-property nature of groundwater resources and aquifer systems, many of the services produced by groundwater are public goods (in a sense, "common property"). While individuals have direct control only over the stream of benefits that flow from irrigating their own lands and meeting their own drinking or other water needs, many groundwater services—environmental maintenance, drought insurance and poverty alleviation—accrue in an aggregate manner to regions or society as a whole. For individuals, the value of these wider social "services" is often intangible. For example, the value of poverty alleviation, while of great importance to society as a whole, may have little relevance for individual landowners. Similarly, individual landowners may be completely unaware of base flows of rivers and streams generated by high groundwater levels and, even if they are aware, these flows may not be of any direct benefit to them. The net result is a discontinuity between the incentives for individuals to maximize benefits accruing to them personally from groundwater utilization, and practices that would maximize benefits for society as a whole.

Multiple communal handpumps in Eastern Region, Ghana, tapping a limited common-property resource in a granitic basement complex aquifer .

5.1.1 Market failures

Groundwater tends to be undervalued in general. The *in situ* values of groundwater, and, from the perspective of this publication, many of the social values associated with groundwater use as well, are poorly reflected in market transactions (National Research Council, 1997, p. 1). As a consequence, from an economic perspective, market mechanisms tend to "mis-allocate" water. Box 20 addresses some of the current issues in groundwater valuation. In any event, efficient markets require clear, enforceable (and often private) property rights, high levels of information availability, minimal externalities, low transaction costs, feasible competition, absence of common-good characteristics, and clear and definite "solutions of continuity". The latter are particularly important in respect to groundwater, since there are rarely single users drawing upon discrete aquifers or hydraulically bounded portions of an aquifer. Since few of these circumstances apply in the case of groundwater, regulation will generally be required to balance equity, environmental and other public interest concerns.

Trading in pumped groundwater is, however, widespread in South Asia (Shah, 1995; Moench, 1994c; Meinzen-Dick, 1996) and a broad range of other developed and developing countries as well. Unregulated groundwater markets create strong incentives for over-extraction. This is well documented in the case of informal markets for groundwater irrigation in developing countries (Shah, 1995) and implicit in the general undervaluation of groundwater resources documented in the United States (National Research Council, 1997). However, the common-property nature of groundwater resources, and the fact that many of the services they produce are public goods, result in failure of the market as an effective mechanism for ensuring efficient and equitable allocation and use (see box 21).

Box 20 Issues in groundwater valuation

Estimates of the value of groundwater have assumed great importance in the context of the massive investments required to avoid pollution, remediate polluted aquifers and control overdraught. Some objective basis, related to the total economic value of the resource (including wider social and environmental values), is required to determine how much to invest in specific situations. Unfortunately, available valuation methods are, at best, partial.

Methods for quantifying the economic value of groundwater resources include: (1) contingent valuation—which essentially involves asking people how much they would pay to maintain the resource or services dependent on it under carefully specified conditions; (2) hedonic pricing—e.g., obtaining a measure of the value of groundwater through differences in the value of lands with and without access to it; (3) derived demand and production cost analysis—essentially, estimating the contribution of water to profits within a given set of economic activities; (4) loss analysis—estimating the value of groundwater as equivalent to the total social costs incurred when drought or depletion constrains economic activity; (5) averting behaviour in which the value of groundwater is estimated by the investments made to avoid water shortage; and (6) substitution—the value of groundwater as equivalent to the least-cost alternative source of supply for meeting the same set of services. Many of the above methodologies are discussed in detail in the specific context of the United States (National Research Council, 1997) and elsewhere (Schiffler, 1998).

The above quantitative techniques do not adequately capture many of the qualitative values associated with groundwater. They also assume perfect knowledge of the groundwater system in space and time and the precise impact of abstractions and uses (including disposal). Non-use and *in situ* values, such as the prevention of saline intrusion and protection of the environment against low-frequency events, are particularly difficult to establish. Valuation of ecosystem losses due to changes in groundwater conditions is, for example, an inherently subjective exercise. Some sections of society approach the environment from a spiritual and religious perspective and, in doing so, place essentially infinite value on the natural resource base that sustains environmental conditions. Other sections of society focus much more heavily on specific economic activities when valuing natural resources. Weighing the relative importance to give to these different perspectives is in itself a subjective value judgement—and one that is often implicit in the valuation methodology selected. A similar ethical divide surrounds questions of ability versus willingness to pay. Access to groundwater is often a critical factor in the quality of life for socio-economically marginal populations in developing countries. Their ability to pay for maintaining groundwater resources is, however, often minimal. How this contradiction can be resolved depends on whether groundwater is treated as a common heritage—to which all have equal and fundamental rights—or whether it is treated as an economic good.

Beyond the above ethical debates lie practical questions regarding the degree of social and technical understanding of groundwater on which value estimates are established. A basic problem has to do with the interlinked and sequential nature of groundwater uses. Often, "waste" from one use is the primary source of water for a subsequent use. Tracing the chain of uses is a difficult task in itself, to say nothing of valuing the contribution of groundwater to each one. Beyond this, it is often difficult to determine whether values adequately reflect low-frequency but high-consequence events. This is closely related to issues of information and understanding. Groundwater is often viewed, by the general public, as a quasi-static stock—"an underground bowl regularly replenished by rain". Values derived from this viewpoint are unlikely to reflect the finite nature of many groundwater resources or their vulnerability to pollution. Similarly, it is difficult to determine how much other forms of capital can substitute for groundwater should existing uses constrain supply availability. The limits of substitution may, after all, only become evident when a chain of sequential and interdependent uses is fully understood. In sum, how well people understand groundwater and its contributions to economic, environmental and other services may well influence the value they place on the resource.

Most services that accrue to individuals are extractive—i.e., they are generated by extracting groundwater and using it for some specific productive purpose, such as drinking water or irrigation. The total economic value of groundwater is, however, the sum of its extractive and *in situ* values (National Research Council, 1997). Most *in situ* values stem from services that accrue to society as a whole and derive from both the groundwater and the aquifer systems. As they are often intangible public goods, the value of these *in situ* services is not fully reflected in the value individual users attach to groundwater or its selling price where water markets exist. It is also rarely reflected in the tariff structure that organizations responsible for managing groundwater resources typically impose, where such organizations exist. The net result is that extractive services tend to dominate individual decision-making, while *in situ* services are undervalued, since there is no clear economic signal given to users for these services. Put more directly, the common-property nature of the resource and the ignorance of the *in situ* services create strong incentives for extraction and few incentives for conservation, particularly in areas where groundwater resources are limited. Individuals have strong incentives to pump as much as they can use and few incentives to invest in managing the resource base.

Externalities associated with groundwater use are another factor underlying market failures. Within any hydrologic or hydrogeologic system, uses are generally sequential and interdependent. Irrigation return flows, for example, can become the primary source of water for wetlands. These can, in turn, be an important source of groundwater recharge, but also of pollution. Quality or quantity changes at any point in the chain of uses affect quality and quantity at subsequent points in the chain. Externalities of this type are not reflected in the value of groundwater to individual users, but do affect both the total economic value of groundwater services and the way benefits are distributed. Externalities generate major equity concerns. Water-level declines, for example, may not reduce the volume of agricultural products produced through groundwater irrigation. If, however, the declines exclude poor farmers (those unable to afford the cost of well deepening) from access to groundwater, then wealthy farmers will capture most of the benefits generated through irrigation.

In many parts of the world, groundwater rights systems have been developed in response to problems stemming from the "common-property" characteristics of groundwater resources and the externalities associated with their use. Generally, rights systems are exclusionary mechanisms designed to allocate shares of the available resource to specific users and/or uses. They are also often designed to limit or eliminate externalities. Here a distinction needs to be made between use rights and the rights to groundwater usually accorded to property owners. The latter is the least complicated (and most widespread) form of groundwater right. Landowners are able to drill or dig wells on their own land and use the water on their lands. Non-landowners are excluded from access to the resource. More complicated systems based, for example, on estimates of the physical productivity of aquifers, the history of past use and the amount of water actually "consumed" in different applications have also evolved in some locations, such as the western United States. Development of these types of rights systems is often advocated as a key factor enabling the evolution of water markets as major mechanisms for water allocation and management (Rosegrant and Schleyer, 1994).

Box 21 Groundwater markets: the differing roles

Misconceptions regarding water markets and the role of rights systems have led to much confusion in debates over groundwater management. Western perspectives tend to reflect the recent development of regulated water markets in locations such as the western United States. These regulated markets are evolving in the context of formal water rights systems. They are widely advocated as mechanisms enabling reallocation of available water supplies to high-valued uses and encouraging water-use efficiency (Reisner and Bates, 1990, p. 130; National Research Council, 1992, p. 2).

In contrast, much of the literature on water markets in developing countries focuses on informal markets that have emerged for irrigation and drinking water. These markets are generally unregulated and operate in the absence of any formal rights system. In urban areas of developing countries, informal water markets are often a key mechanism for all sectors of society to obtain access to supplies for domestic use and, in the process, serve to reallocate water from agriculture to higher-value uses. Most literature on informal water markets, however, emphasizes the role they play as mechanisms enabling resource-poor farmers (those unable to afford the cost of wells) to obtain access to irrigation water (Shah, 1993). By expanding groundwater irrigation, these have enabled resource-poor farmers to increase their incomes and productivity. The depth and breadth of informal markets are important to recognize. According to Shah, in most of India over half of the annual pumping hours are probably used for selling irrigation water, and the area irrigated using purchased water is probably as large as the total area irrigated from surface sources. The literature on informal markets also emphasizes the incentives they create for over-extraction of groundwater in the absence of regulated rights systems (Moench, 1995b; Shah, 1993). Views on the benefits of informal water markets are not universal. In some cases, water sales can become mechanisms for large farmers to control their less wealthy neighbours (Janakarajan, 1994).

If regulated markets are to evolve into major mechanisms for groundwater allocation and management in developing countries, water rights systems will need to be established. Formal markets based on formal rights systems can address some of the management issues stemming from the common-pool nature of groundwater resources. Attention should not, however, be focused only on these formal structures. Informal markets are a core feature of water use in many developing countries. They underlie much of the food security and poverty alleviation benefits derived from groundwater resources. Government policy instruments, particularly those related to energy supply and pricing, influence the structure and working of informal markets. Research in India indicates that subsidized power, particularly if supplied at a flat annual rate, can have equity benefits by encouraging well owners to sell water at low rates to those who do not own wells. This, however, greatly exacerbates tendencies to overdraw resources where groundwater is limited but could be beneficial in areas suffering from rising water levels or waterlogging. Overall, it is important for governments to develop a cogent understanding of how informal markets function and to develop policy instruments through which they can influence the behaviour of millions of private pump owners and their customers.

The details of different groundwater rights systems or the role of water markets as management and allocation mechanisms cannot be covered in detail here.[10] Three points are, however, important to make. First, rights systems that capture fully the public good characteristics of many of the services produced by groundwater have not been developed. As a result, the value of these services is not reflected in market prices for such rights. Therefore, markets alone cannot be viewed as adequate mechanisms for groundwater allocation. Second, rights systems in general are unable to address externalities associated with groundwater use. Concepts such as "consumptive use" (the amount of water actually consumed by a given application) have been developed in attempts to limit externalities—but success is, at best, partial. Third, the development of groundwater rights systems is often constrained by substantial practical hurdles. Taking the case of India and China, each has millions of private wells and the registration of each individual user is unlikely to be practical. Monitoring and enforcement of rights based on volumetric use for even a small fraction of them would pose large, if not insurmountable, logistical difficulties. The net result is that the development of groundwater rights systems, however refined these might be, is not sufficient to enable markets by themselves to function as an efficient or equitable mechanism for groundwater allocation.

Traditional mechanisms that exist for allocating surface water, such as this flow divider in Nepal, are difficult to apply in the case of groundwater—where there are no precise "solutions of continuity".

[10] See Caponera (1992) for a detailed discussion of legal structures.

In conclusion, markets represent one dimension in the institutional framework for groundwater allocation and management. Given the common-property (often open-access) nature of groundwater resources, markets do not capture many of the social, environmental and other third-party implications of allocation decisions. Governmental, private or cooperative institutions for resource management are, as a result, essential.

5.1.2 Institutional failures

The common-property, or "invisible", nature of groundwater resources, combined with the often intangible nature of the services they provide, creates particular challenges for the development of effective management institutions. The extensive literature on common property indicates that the feasibility of establishing institutions capable of effective management is influenced by a number of factors including:

- Clearly defined resource boundaries;
- Adequate information on resource conditions and dynamics;
- A clearly defined (and generally small) group of users;
- Ability to monitor resource use;
- Ability to control and exclude free-riders (individuals who receive benefits from management but do not cooperate with the actions needed to achieve it).[11]

The above list is not intended to be exhaustive. It is, however, important to recognize the complexity of developing effective groundwater management institutions.

Aquifer boundaries pose problems for institutions and administration. There is generally non-conformance between the groundwater flow domain and the essentially "public" domain in which decisions are made about the allocation and protection of groundwater. The flow domain does not have clear, steady boundaries which match the generally precise administrative and jurisdictional bounds in the public domain—the so-called "solutions of continuity". While hydrogeologists are always seeking to define physical boundary conditions, these natural solutions of continuity are dynamic—changing over space and time—and rarely monitored.

Information availability also represents a key challenge to the development of management institutions. Establishment of an effective institution for water resource management is much more straightforward when sufficient data are available to convince key policy and user audiences of the importance and viability of management. If data are insufficient to document the nature of emerging problems or to understand aquifer system dynamics, then management objectives and options will also remain unclear. In this case, the task of convincing local users and policy actors to support or participate in management becomes harder. Furthermore, management approaches based on insufficient understanding of aquifer systems often fail. This can undermine the overall credibility of the management institution.

Because of uncertainties in defining aquifer system boundaries, it may also be difficult to distinguish user groups. Perhaps more importantly, aquifer systems often have large numbers of users, particularly where landholdings are small.

---

[11] See BOSTID (1986) and Ciriacy-Wantrup and Bishop (1975).

Successful negotiations over aquifer management in the western United States generally involve no more than a few dozen active participants—large family-owned or corporate agribusiness interests, municipal water-supply entities, Indian tribes, environmental organizations, water districts and government departments. Actual rights holders may run into the hundreds but are generally represented by larger entities. Furthermore, most of the large entities are relatively sophisticated—they often have their own technical and legal staff and have the ability to analyse scientific data on aquifers. In contrast, in many developing countries the number of individual groundwater users is far larger. No intermediate organizations exist to represent the views and interests of individual users and the users themselves lack technical capabilities.

A Ground Water Department monitoring borehole in Rajasthan

Here most monitoring is manual and only occurs once or twice a year.

Monitoring groundwater use is much more difficult than monitoring the use of, for example, land or forest resources. But the ability to monitor resource use is often critical for the development of common-property management institutions since, unless use can be monitored, individuals have little assurance that other users are cooperating with management decisions. Metering provides the highest level of assurance regarding extraction. Large numbers of users and dispersed use locations can, however, make metering both difficult and expensive. In many countries, groundwater extraction is monitored through indirect measures—such as pump electricity consumption or the size of irrigated areas. Even these indirect measures can be problematic. In India, for example, electricity meters on agricultural pumps were eliminated in the mid-1980s because of conflicts between farmers and meter readers and metering costs. At present, most agricultural electricity consumption is not metered and there is substantial political opposition from farmers to any return to metered systems. Monitoring is, thus, a major challenge facing development of groundwater management institutions.

Finally, the ability to control free-riders is commonly emphasized in the common-property literature as essential for effective management. In the case of groundwater, such management activities as the adoption of efficient irrigation technologies and agreed-on reductions are often essential to address emerging problems. Ensuring that all individuals within a management area comply with management decisions of this type will be difficult. Wells are generally located on private lands and management traditions are generally lacking. As a result, control over free-riders complicates the success of management institutions.

5.2 Policy failures and the impact of success

Policy history represents a major root cause of competition and emerging groundwater problems. Historically, most governmental policies enabled groundwater development as a way to settle populations and develop agriculture. These policies have been successful only to the extent that governments now face the difficult task of reforming policy frameworks to reflect needs for managing groundwater demand. The magnitude of this task should not be underestimated. In many countries, irrigation development has been a major focus of social organization and state legitimization for centuries if not longer (Wittfogel, 1957). The policy history of large-scale groundwater development is much more recent, but it fits tightly into long-established historical frameworks.

Until the last three decades, in most countries groundwater resources were viewed as essentially unlimited. Overdraught and pollution were not concerns because the resource was "infinite"—or at least relatively undeveloped. The fact that under many legal systems groundwater is tied to landownership goes with a worldview that sees land rather than groundwater as a limiting factor. Beyond this, many countries have developed policy frameworks explicitly intended to encourage economic expansion through unlimited groundwater development. The case of India is very revealing. In the post-independence period, India's development policies have emphasized food production through dissemination of the green revolution package of agricultural technologies. Irrigation is the lead input required for this package to be effective. As a result, expansion of irrigated areas has been a major focus in development plans from the 1960s to the present. Where groundwater is concerned, two primary policy tools were used to achieve this objective: subsidized credit for installation of energized wells through the National Bank for Agriculture and Rural Development (NABARD), and energy subsidies. Subsidies for drilling wells and purchasing pumps are available in rural areas except where groundwater is classified as overdeveloped. To complement this, major efforts were made to expand the electricity system into rural areas for the purpose of groundwater pumping. Furthermore, electricity and, to a lesser extent, diesel are provided at subsidized rates to agriculture, policies which continue to the present day. Electrical power is provided free of charge for pumping in some Indian states. As a result, for example in Haryana, official figures indicate that over 50 per cent of electricity consumption is in the agricultural sector—and most of this is consumed by groundwater pumping. Furthermore, overdraught problems are exacerbated in many sections of Haryana where groundwater quality is high.

The above case of India is not unique. In many developing countries, agricultural subsidies, encouraging excessive groundwater use, have created major groundwater problems. In Mendoza Province, Argentina, for example, agricultural subsidies for irrigation (below-market interest rates on public loans, tax exemptions, and guaranteed purchase of produce at a minimum price) caused the depletion and salinization of aquifers. Interestingly, this was a case of good information promoting a bad policy: after a technical assessment showed groundwater availability, agricultural interests lobbied at the national and state levels to secure special concessions, including low interest rates, public loans, tax exemptions and guaranteed purchases and prices for produce. The success of these policies resulted in the deterioration of the aquifer and ultimately bankruptcy of the regional economy. In this case, unbalanced stakeholder participation resulted in a policy favouring a special interest group (large farmers, well drillers and agribusiness), with ultimately negative results for the resource base and for the economy as a whole.

In general, it is important to recognize that many groundwater degradation and competition problems are the direct result of policies that encourage unlimited exploitation. The popularity of these policies makes it very difficult for governments to convince constituents of the need to move from development to management.

Private drilling for high-discharge centre pivot irrigation systems in the Batinah coastal plain, Oman

Controls on private groundwater abstraction have proved difficult to implement.

5.3 Perceptions of groundwater: problems of scale, distribution and time

Public and policy-maker perceptions of groundwater represent another important root cause of emerging problems. In many cases, regardless of the degree of formal education individuals have had, perceptions of groundwater resource dynamics are partial at best. Groundwater is often viewed, for example, as an inexhaustible resource, cleaned by the filtering action of aquifers and held, as in a "bowl" or "lake", or "underground river". These perceptions do not reflect reality, and often result in use patterns that cause unanticipated problems. Most misunderstandings relate to the scale of aquifer systems, the distribution of groundwater within them and the timescales on which groundwater systems function (see figure 7).

Structural features that determine aquifer scale or the distribution of groundwater within anisotropic fracture or sedimentary structures tend to be poorly understood. Aquifer systems range in scale from highly localized ones covering an area of a few hectares to the massive regional aquifers of sedimentary basins underlying many thousand square kilometres. Groundwater flow within aquifer systems is continuous and the overall balance between extraction and inflow is often seen as the primary factor influencing the sustainability of use patterns. This said, however, groundwater conditions at any given point within an aquifer system are more heavily influenced by specific local conditions than by the overall status of the system. Water levels, for example, often reflect drawdown attributable to pumping within a local well field, recharge from an adjacent river or lake, or the presence of local low-permeability layers. These localized effects are time-dependent and can have a major impact on the tangible conditions—pumping costs, drainage, wetland distribution etc.—that users face.

The actual pattern of groundwater flow and storage conditions within aquifer systems is another factor complicating perceptions. This is particularly true in regions underlain by metamorphosed crystalline rock, or basement complex, where groundwater flow can be heavily influenced by fracture patterns that often do not conform to watershed boundaries or other surface features. Even in "simple" sedimentary basins, however, changes in the nature of sediments greatly influence groundwater characteristics at specific locations. Lateral permeability is often much greater than vertical. Sand or gravel beds often function as major pathways for rapid lateral flow and, as a result, serve as sites for both high-yielding wells and rapid transmission of pollutants, while more clayey layers restrict vertical flow. Thin alluvial aquifers may be pumped perennially if continuously recharged by an adjacent watercourse. In contrast, basement complex aquifers can be rapidly drained at the limit of base-flow recession where specific yield of the aquifer tends to zero. The characteristics of large multilayered aquifers often change with development as vertical leakage between semi-permeable layers and compaction alters the hydraulic conductivity and storage characteristics of the sediments and new flow pathways are created by penetrating wells. These factors can significantly alter both recharge characteristics and the migration of low-quality or polluted water. This, "conditioning" of the groundwater resource by modes of groundwater abstraction is largely ignored in most contemporary development strategies and has resulted in major adverse impacts, such as large-scale salinization and subsidence. Thus, management approaches based on the perception of aquifers as relatively homogeneous and static systems tend to miss fundamental aquifer dynamics.

Figure 7 Contrasting flow regimes in aquifer systems

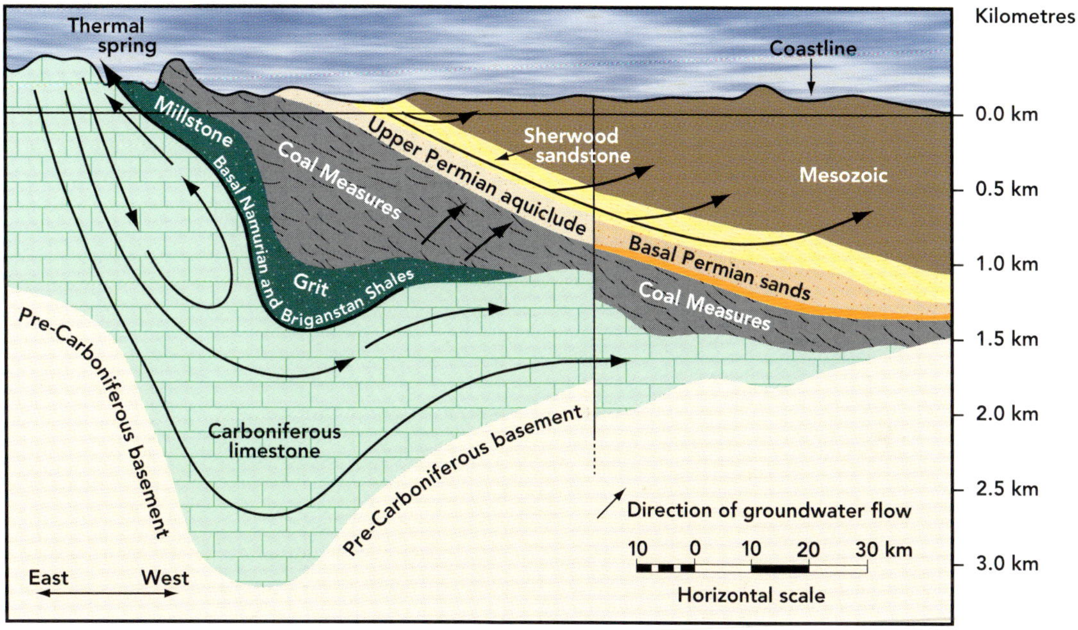

Schematic cross-section through the East Midlands, United Kingdom, showing principal directions of groundwater flow
Source: Downing et al. (1987)

Finally, there is the question of time. The dynamics of aquifer systems are different from, and often much more conservative than, those of surface-water systems. In shallow systems, groundwater recharge and discharge occur rapidly and are closely linked to rainfall and the flows in surface streams. This is particularly true, for example, in karst topography, where cavernous limestone provides direct links between surface-water and groundwater systems. For many shallow discontinuous alluvial and basement complex aquifers, groundwater flow is a function of annual recharge and base-flow recession (Wright and Burgess, 1992; Herbert et al., 1996). In other hydrogeologic contexts, however, groundwater-flow dynamics commonly operate on much longer timescales. Flows occurring in large sedimentary basins can be governed by paleo-recharge events that occurred thousands of years ago in addition to thermal and density contrasts that can be expected at depth (Bachu and Underschultz, 1992). As a result, groundwater flows and reserves can be either closely related or essentially unrelated to contemporary rainfall patterns, but perceptions about the sustainable use of the resource rarely take this into account.

The above misconceptions are a major factor contributing to the emergence of groundwater problems and competition over the resource. Until very recently in most countries, policy frameworks and public discourse regarding groundwater focused on exploration and development issues. The expansion of borehole and handpump installations in sub-Saharan Africa and elsewhere during the International Drinking Water Supply and Sanitation Decade (1981-1990) is a case in point. Here the focus was on the installation of wells and boreholes equipped with handpumps. Results were mixed (World Bank, 1996) because management needs have remained outside the dialogue, in large part because the resource was poorly understood. This is now gradually changing in some regions where the limits of the resource base are better understood. The case of the High Plains, Ogallala, aquifer in Colorado, Kansas, Nebraska and Texas is illustrative. Dr. David Kromm, an academic involved in groundwater management research at Kansas State University, comments that when he began his studies on groundwater organizations in the western United States over 20 years ago, farmers often described the Ogallala aquifer as an underground lake or stream. Since then, understanding has improved as farm communities have been forced by declining water levels and groundwater contamination to recognize the nature of groundwater systems and the limitations on use imposed by aquifer characteristics. Now concepts such as aquifer permeability and the effects of semi-confining layers (aquitards) are relatively widespread among farmers and others whose livelihoods involve direct use of groundwater.[12] Even at present, however, groups of individuals seeking to form groundwater protection committees in the High Plains often invest first in programmes to educate themselves concerning basic hydrogeologic concepts and the specific nature of groundwater resources in their area.[13] These groups are self-selected on the basis of personal interest and concern regarding groundwater problems in their areas. Their decision to invest first in their own education reflects both their own awareness regarding the complexity of groundwater resource dynamics and the limited nature of understanding of groundwater resources, even among those directly concerned with initiating management.

Hence, the limited understanding within both user and policy communities regarding the nature of groundwater resources and hydrogeologic system dynamics represents one of the primary reasons potential management needs have remained unidentified and simply not addressed. Even now, as understanding begins to spread among those most directly concerned with groundwater problems, education is a critical need.

5.4 Lack of public awareness

Closely linked with lack of understanding of the resource is lack of awareness on the part of the public and policy makers regarding the importance of the resource, the significance of emerging problems and the nature of management options. Given the highly diversified and often localized nature of many groundwater problems and the importance of stakeholder involvement in the development of solutions, public awareness is another essential need.

---

[12] Personal communication to M. Moench, 1 November 1998.

[13] Moench, interviews with the directors of groundwater districts in Colorado, Kansas, Nebraska and Texas between 1993 and 1998.

The general public tends to be unaware of the role that groundwater plays as a fundamental input to highly valued services, such as ensuring food security, protecting the environment and maintaining a reliable source of drinking-water supply. Groundwater is "invisible" and ubiquitous. Because it is a common-property resource, individuals have little reason for monitoring availability or quality except as those factors directly relate to their own specific uses or needs. In addition, scientific information has generally been regarded as necessary only to the technical community. Public and policy-maker misconceptions regarding groundwater are, as a result, common. These misconceptions contribute to the common undervaluation of groundwater resources (see box 20).

Groundwater recharge message, Gujurat

While education is essential for the development of management systems, it must be carefully targeted and developed to be effective. Individual water users are often keenly aware of problems that directly affect the activities on which they depend. Some farmers, for example, can provide more comprehensive information on groundwater quality changes and fluctuation patterns than is available from state groundwater monitoring wells. At the same time, the level of understanding of regional hydrogeologic dynamics among farmers and other local water users is highly variable. Some users can provide detailed explanations based on local geologic features for the movement of groundwater resources, while for others groundwater is viewed as part of an endless supply for which they only need to dig deeper to maintain access.

The core point here is that local knowledge of groundwater systems is highly variable. Educational materials that repeat knowledge that most users already possess will be of little use. On the other hand, misconceptions need to be addressed and practical avenues for addressing specific problems explained. Furthermore, since local users often have a far greater understanding of social (and in some cases hydrogeologic) features within their local areas than outsiders have, education cannot be viewed as a one-way process. Education needs to be viewed as a process through which stakeholders *and* groundwater specialists are able to arrive at a common understanding regarding management needs and options.

Ibb, Yemen

Intense local precipitation may not contribute much recharge.

5.5 Variations in resource characteristics, social conditions and management options

Responses to emerging groundwater problems are complicated by variations in resource characteristics and social conditions. Such variations limit the ability of governments or other entities to develop standardized approaches to management. Three types of variation have major implications for groundwater management: hydrogeologic, climatic and social.

Hydrogeologic variation relates to the great changes in groundwater resource dynamics that exist both between and within aquifer systems. Many of the factors related to this were discussed previously in section 5.3. It is essential to recognize, however, that the great natural variation in groundwater conditions affects actual management needs and options as well as perceptions. In many cases, variation between areas necessitates highly localized approaches to management; in other cases, regionally based approaches are sufficient.

Natural variation in climatic conditions is also important since precipitation characteristics greatly influence management options. The ability to capture run-off for recharge of aquifers, for example, depends heavily on the intensity and duration of precipitation events. In many parts of the world, estimates of the amount of water available for direct use or recharge are based on monthly or even annual averages. In locations where rainfall events are either seasonal or of brief duration and high intensity, this can be extremely misleading. Take the case of Kutch district in Gujarat on the India-Pakistan border. The area near Mandvi, which has an average annual rainfall of 350 mm, received 654 mm over a four-day period in July 1992 after receiving only 185 mm total in 1991.[14] Half the annual rainfall in Kutch typically occurs over a period of 2-3 hours during the monsoon season. There are generally only 8-10 rainy days in the year and rain actually falls for an annual average of 12-15 hours (Pisharote, 1992). Under these conditions, run-off is intense and lasts for brief periods. Even in high-rainfall areas of Gujarat, precipitation is highly seasonal. Out of an annual average of 51 rainy days in the south, 48.5 (accounting for 94 per cent of the total rainfall) occur from June to September (Phadtare, 1988).[15] The above conditions are common in many arid regions and contrast fundamentally with conditions in regions, such as Europe, where precipitation is more evenly distributed. Management approaches need to respond to this variation by incorporating near-real-time approaches to recharge assessment and systematic analysis of recharge styles.

Social variation is often as great as hydrogeologic or climatic variation and may present even greater challenges to the development of groundwater management systems. In the United States, for example, water rights systems vary greatly between states. This reflects different histories, community characteristics, traditions and water management needs. In many rural areas throughout the world, different communities use water in different ways. Pastoral groups relying on animal husbandry for their income have very different sets of water needs from communities engaged in irrigated agriculture. Urban areas differ from adjacent rural areas in both demographic and sociologic make-up and institutional responses to water use. In rural areas, where traditional codes and customs prevail, regional

---

[14] Personal communication, K. C. B. Raju, Director, Central Ground Water Board (Retired).
[15] Calculated from data in Phadtare (1988, p. 7).

Box 22 Conceptual studies and monitoring: the foundation of sustainability

Conceptual studies and monitoring are the foundations that support sustainable development. Determination that a particular rate of groundwater withdrawal or general management plan is sustainable depends on in-depth understanding of the groundwater system. This understanding begins with knowledge of basic hydrogeological processes and impacts—recharge, discharge, the location of potential sources of contamination etc. This understanding may lead to the development of a conceptual model. Such a model should be augmented by a programme of monitoring to provide the data essential to generate a quantitative perspective on the status of the groundwater system and to validate the conceptual understanding.

The practical importance of conceptual models and monitoring is illustrated by the case of overdraught and water conservation in agriculture. Some experts view improvements in water-use efficiency as a key mechanism in order to reduce overall water demand, while other experts view savings achieved at the field level as making little difference at the level of hydrologic systems. Both sides of this debate agree that critical data are lacking to determine the relative balance between waste and savings under different conditions (Moore and Seckler, 1985). Resolution of this conceptual debate has major implications for the massive investments in irrigation efficiency currently being made by individual farmers and national governments. Where water savings are real, investments are probably justified. Where water conservation at the field level simply reduces return flows that serve as sources of supply for subsequent users, the economic justification for investments is far less clear. Monitoring is also essential in order to determine the need for agricultural water conservation at a stage when management can be effective. Problems such as groundwater overdraught often emerge gradually. If they are identified and actions initiated to address them early, management may cause little social disruption. If, on the other hand, responses are delayed until major problems arise, massive social disruption may be unavoidable. In Yemen, for example, groundwater extraction in some locations is 300 to 400 per cent of recharge and water levels are falling by 2-3 metres a year. To return groundwater use to sustainable levels, extraction needs to be reduced to a quarter of current levels. The social disruption that would occur if such reductions were imposed by the Government would be politically difficult as well as socially unacceptable. If, however, water levels had been monitored early and the data used to develop conservation programmes before problems became critical, then politically and socially viable solutions might have been possible to develop.

From a broader perspective, the development of a conceptual model and monitoring programme enables interpretation of present and past conditions. Projecting future conditions is analytically difficult and must build upon an understanding of present conditions. Planning requires making predictive analyses and generally involves the use of computer modelling techniques. The conceptual understanding of the groundwater system provides the basis for the proper construction of the computer model.

Determining the sustainability of a water supply with any degree of confidence can be done only in a planning environment having detailed information and understanding. Ultimately, though, water-supply development policy involves trade-offs. Most hydrologic systems have ecosystems, landscape elements, or pre-existing water users that are dependent on current discharge or recharge patterns. Further development may require trading off these dependencies in favour of new water-supply plans or policy. If dependencies are not well understood or considered, management changes may have major unanticipated impacts.

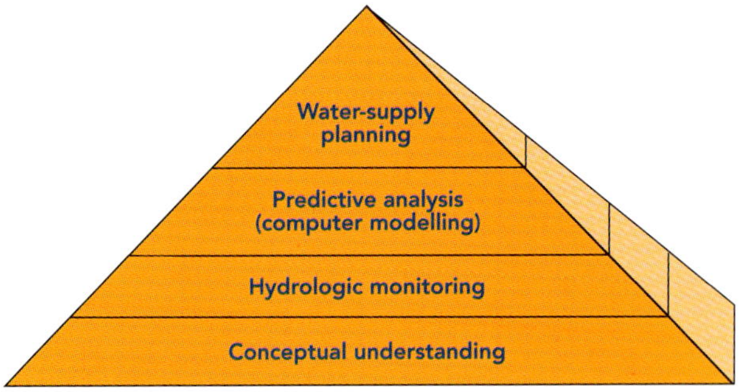

institutions representing classes of water users are often weak or absent. Urban areas, in contrast, are generally served by municipal water-supply organizations that represent a large class of users. In the urban case, water management initiatives may be able to draw on existing institutional structures. In rural areas, similar initiatives may need to focus far more heavily on individual users.

The different types of variation have fundamental implications for groundwater management. Institutions and approaches need to be closely attuned to local conditions. Flexibility is essential here, since solutions that work in one area may be inappropriate in adjacent areas. This creates major challenges for the development

Activities to caputure water for agricultural activities depend on individual farmers but can be highly effective.

of management institutions. Solutions based on models developed in pilot management areas cannot easily be replicated in other areas. Centralized approaches are often confounded by the great variation in conditions, problems and management options that characterize local areas. This reduces the ability of governments to manage groundwater resources directly through the apparatus of the State. It also emphasizes the importance of involving local communities (those who are most familiar with local conditions and most directly affected by groundwater problems) in the management of groundwater resources.

This having been said, and taking note of the legal responses to groundwater management outlined in section 5.1, care has to be taken to ensure that all public interest matters are invested somewhere. Stakeholder participation alone is no guarantee that wider public interest matters will be addressed, since stakeholders invariably have a degree of self-interest (Kemper, 1996). There is also the need to ensure that patterns of local consumption do not impose onerous externalities in the framework of the aquifer system as a whole.

5.6 Limited management capacities

Limited technical and administrative capacities in local communities and government institutions at all levels place critical constraints on the development of groundwater management institutions and approaches. The relative scarcity of hydrogeologists in many developing countries is one factor contributing to this situation. Deeper factors, however, relate to the history of groundwater development, traditions and cultural characteristics, and use patterns.

- ***Development history.*** In most developing countries, intensive groundwater development is a phenomenon of the last few decades. Most government organizations concerned with groundwater were formed primarily to provide technical support services, such as groundwater exploration and well drilling, to support development. Only recently have groundwater overdraught, quality degradation and pollution emerged as concerns. As a result, few groundwater organizations were designed to address such concerns or have much experience in doing so. Only in the relatively recent past have these organizations expanded their activities to include management of the resource base.
- ***Traditions and cultural characteristics.*** The willingness of different communities to participate in groundwater management may depend heavily on cultural characteristics. Some regions and countries have long traditions of centralized control over a wide variety of aspects of life. In such regions, socially accepted administrative mechanisms may exist that enable direct regulation of individual water use. In other regions, centralized control traditions are absent and groundwater use is a deeply entrenched traditional right of landowners. Governments and local authorities and organizations operating in this context often lack the capacity to implement regulatory approaches to management. In some cases, such as the High Plains Underground Water Conservation District in Texas, the inability to regulate use has led to the development of management approaches relying heavily on education (Moench, 1991). Such approaches, however, require outreach and educational capacities that are often absent within governments and local societies of developing countries.
- ***User numbers.*** The number of groundwater users is often a critical factor in the capacity of societies to develop management institutions. For example, differences between the United States and the South Asian countries Bangladesh, India and Pakistan have been highlighted earlier. Government capacities are not just limited in instances such as South Asia, where extremely large numbers of private and public wells are common. In France, even relatively simple activities, such as the installation of meters for monitoring extraction by irrigators, are, in some cases, a source of conflict (Barraqué, 1997a). As Dr. Poitrinal, a member of the ad hoc expert committee, commented, the installation of meters on irrigation wells in France will not be a fast or easy task, although the Ministry of the Environment and basin agencies have been demanding it, and in some cases local associations support the process. While these facts may render some solutions impractical, it is clearly prudent to act while the resource base can still be protected.

Box 23 Conjunctive management in California

In California, a drive is under way to forge a consensus among disparate water interests. Forums fostering dialogue among stakeholder groups offer a unique opportunity for exploring new water management options, such as conjunctive management of surface water and groundwater. The integration of storage capacity currently available in California's overdrawn aquifers into the state's water system is emerging as a viable alternative to the construction of new major on-stream storage facilities. Unfortunately, the ultimate integration of conjunctive use into California's water system is hampered by a comparative lack of experience with this storage mode and by the degree of uncertainty surrounding its storage enhancement potential. Hydrologic analysis conducted by the Natural Heritage Institute suggested that the integration of conjunctive use in the Central Valley water system could significantly increase supply, although a large set of very complex technical and institutional challenges arise when such an integration is considered.

A desire to surmount these challenges alerted a variety of stakeholders to the need for an analytical tool to account for the yield generated through integration of conjunctive use into the California water system. These stakeholders determined that this tool should be: comprehensive, in that it includes all aspects of the water system; transparent, in that the data used and assumption implicit in the model are readily apparent; flexible, in that model parameters can be quickly adjusted and the results analysed; and able to compare alternatives, so that scenarios can be compared. The Water Evaluation and Planning System, or WEAP, developed by the Tellus Institute of the Stockholm Environmental Institute—Boston, was chosen because it possesses each of these characteristics. In addition, WEAP includes an easy-to-use graphical user interface which allows the user to include all aspects of a water system: hydrology; infrastructure; demand; and in-stream flow regulation. WEAP was used to investigate the potential benefit of integrating the storage capacity available in local aquifers into a statewide conjunctive-use programme.

The results of the investigation were encouraging. Using the historical hydrologic trace from 1970 to 1992 as input, both water-supply deliveries to one of California's major water users, the State Water Project, and a key environmental indicator, outflow from the Sacramento/San Joaquin delta, were enhanced during the critically dry 1977 water year. The apparent benefit of conjunctive use, and of WEAP as an analytical framework, prompted a number of important actors in California to invest additional resources in model development and analysis. WEAP is now being used to study the integration of a number of aquifers of into the State's water management strategy.

WEAP model results during the critically dry 1977 water year

(solid line with conjunctive use, dashed line without conjunctive use)

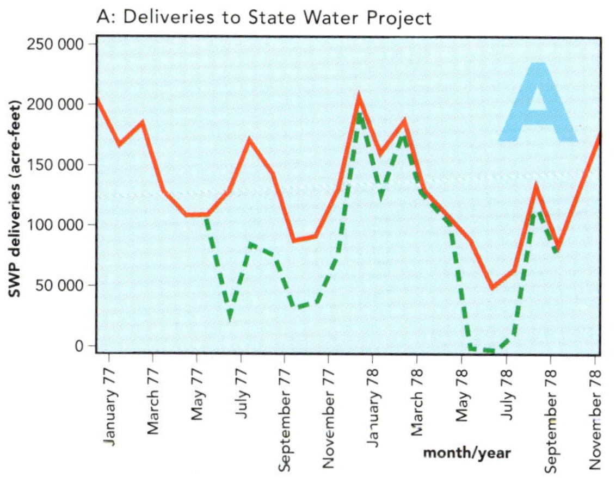

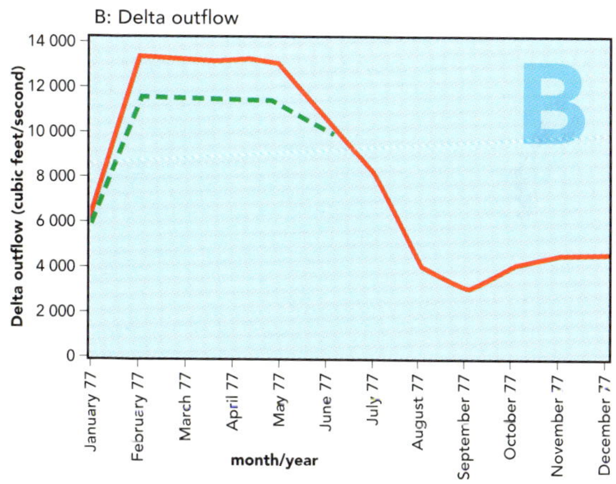

5.7 Lack of data and scientific understanding

Lack of data and scientific understanding of groundwater resources is often a critical gap undermining the development of groundwater management approaches and institutions. The absence of data often limits the degree to which hydrogeologists are able to quantify and describe aquifer dynamics. Aquifer systems are complex. Thus, information on their dynamics is generally essential for management. Equally important, however, are the ways in which raw data and information are treated, presented and used. Information is only useful if it is used. For this to occur, the information must be accessible to potential users and presented in a manner they can understand. With groundwater, problems in explaining complex three-dimensional flows may even confound public debate (Wilson, 1998).

5.7.1 Technical needs for data and information flows

From a technical perspective, as discussed in the section above on perceptions, the absence of information creates situations in which emerging problems and management options are poorly understood. As a result, the essential nature of basic data and research on hydrogeology should be clear. Data collection is, however, expensive and some judgement as to the required precision always has to be exercised. Simply increasing the amount of data collected may provide little assurance that the types of data required for addressing specific problems are available when needed. Generating the technical information required to meet emerging groundwater management needs depends on at least three types of scientific data collection and analytical activities: (1) long-term baseline monitoring; (2) targeted research on basic processes; and (3) site-specific analysis of problems and management options at local levels. Long-term baseline data are essential in order to understand the dynamics of hydrogeologic systems and to provide warning of emerging problems. Research on basic processes is critical for data analysis and interpretation. Detailed studies are essential, because the broad parameters which can be monitored through baseline data collection systems may not be adequate to identify management options that reflect local conditions, even though they should be sufficient to indicate the presence of emerging problems. This is, for example, the case in India, where water-level monitoring programmes indicate long-term declines in some areas but give little insight into the use patterns causing those declines (World Bank, 1998a).

For long-term monitoring, broad categories of data include water-level fluctuations, determination and changes of groundwater flow parameters, water-quality trends and key pollution indicators. This needs to be accompanied by regional analysis of the aquifer systems that will in turn allow targeting of data collection and data types. Emerging management needs identified with long-term monitoring can be the subject of detailed programmes to better understand and characterize local aquifer sytems. Basic scientific research is needed to deal with heterogeneous aquifers, model verification and validation, and relationships between contaminants and aquifer material.

Irrespective of the quality of data and their technical basis, an essential element of any good groundwater information system is the ability for the data and information to flow regularly and reliably between the producers and the users. Given that in comparison to surface-water systems, groundwater systems respond slowly to change, the imperative for real-time or near-real-time data that exists for suface-

Box 24 Identification of options for environmental protection, Tunisia

Examples of ecosystems that depend partly or totally on groundwater are numerous. There is often no inherent conflict between preservation of these ecosystems and regional socio-economic development. In Tunisia, the National Agency for Environment Protection of the Ministry of the Environment has led an effort to protect the Ichkeul National Park with financing by the German Government. This effort shows the possibility of meeting ecological needs while saving water for other uses through carefully targeting water supplies to meet specific ecological goals.

Integrated water resource modelling was combined with identification of the ecological constraints necessary to preserve the equilibrium between existing ecosystems in the Ichkeul lake and related swamps. The most important of these constraints is the seasonal fluctuations in salinity—which range from low levels in winter up to 30-40 grams per litre in summer.

The identified constraints correspond to the environmental conditions necessary for maintaining specific ecological subsystems and the functioning of the lacustrine ecosystem itself. These constraints were based on parameters such as water level, salinity and flow at the lake outlet. For each subsystem, optimum conditions were defined through: (1) ranges of values for specific parameters; (2) the number of months during which a condition is to be fulfilled; and (3) periods in the year during which conditions need to be met. In addition, the frequency of the years in which optimum conditions need to be met was identified, taking into account annual variations in climatic conditions. This is not the same for each ecosystem, since all the years are not equally favourable for each of them.

A 40-year reference period of climatological conditions was selected as representative of the variation of the climatic conditions. The water resource model was run for each stage of water resource development over this reference period. By doing this, it was possible to construct scenarios regarding impacts on the ecosystem associated with different levels of water mobilization. In each scenario, strategies for water release from dams and the management of the gate located on the outlet of the lake were the primary parameters adjusted. Results of the simulations were compared in terms of: (1) gaps between water mobilization objectives and actual amount of water actually available; and (2) fulfilment conditions of the ecological constraints as indexes varying between 0 and 100. In addition to these constraints, indicators have been selected to show eutrophication risk.

Using the above process, a water resource management strategy has been developed that makes possible both tourist activity linked to the preservation of the environment in the lake and swamp and adequate water deliveries for domestic, agricultural and industrial uses.

General sketch of the Ichkeul basin

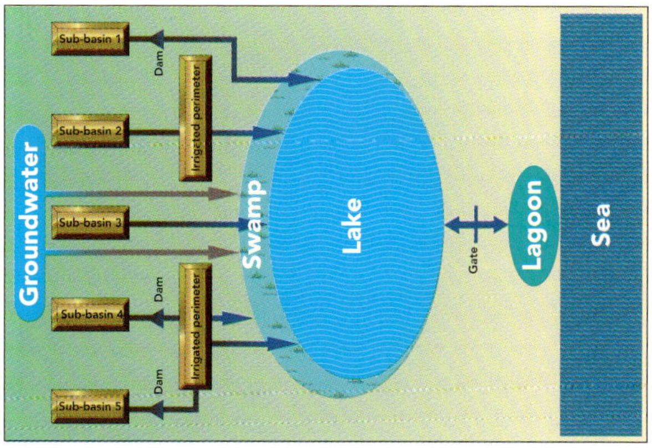

Example of variation of ecological indexes versus water input into the lake

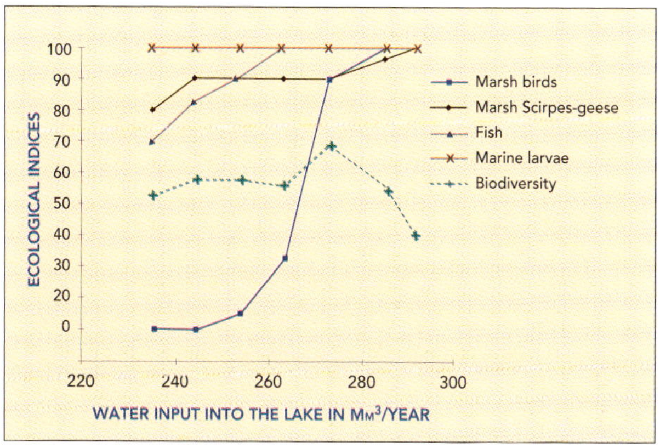

water systems may not apply. Nevertheless, the regular collection, analysis and dissemination of data and information are fundamental for any programme of groundwater management that wants to influence public opinion and policy and decision makers. The role of information is, perhaps, most clear in water management debates in the western United States. There, models and information have become the basic "negotiating texts" groups of stakeholders use to reach agreement on management options (Baker and Romm, 1990). In addition, lack of information and lack of access to information were two of the key points identified by experts and non-governmental organizations in India as constraining the development of effective strategies for managing groundwater (VIKSAT/Pacific Institute, 1993).

5.7.2 Data presentation and use

Technicians often argue strongly for more data—and then when they get them present tentative, difficult-to-interpret results and demand financing to collect more data in order to strengthen those results. Policy makers and the general public, on the other hand, often have unrealistic expectations regarding the ability of technical and scientific analyses to provide straightforward answers, particularly given the general paucity of groundwater data and the limited means with which to gather and analyse them. Hydrogeologic systems are rarely simple. The importance of bridging this gap between the data available and their policy implications relates directly to some of the fundamental challenges facing the development of management institutions for common-property resources.

The willingness of individuals and organizations to participate in the management of common-property resources is strongly influenced by the degree to which they understand resource dynamics, the clarity with which resource boundaries are defined and the ability of management organizations to control free-riders. These factors often depend, in turn, on basic information regarding resource use, conditions and dynamics. Furthermore, because no single entity generally has the power to control use over the objections of existing users, management often requires negotiation among different users and user groups. A key element in many negotiated agreements is the development of a common understanding concerning physical relationships and the probable effects of different management actions. Baker and Romm (1990) point out the value of hydrogeologic models as neutral "negotiating texts" in negotiations over water management in the western United States. As Baker and Romm (1990, p. 6) indicate: "Until the benefits of cooperation can be quantified for each party, it is not feasible to expect a user or user group to invest resources in the cooperative effort." Finally, basic scientific information and data are essential for any groundwater rights system to work. This again relates back to the question of assurance. The value of rights depends directly on the assurance that they actually relate to physical resource characteristics.

Scientific data on groundwater conditions and availability are rarely analysed or presented in ways specifically designed to meet society's information needs for common-property management. Data and the numerical models prepared with them are often regarded primarily as technical inputs necessary for scientists, engineers and planners to develop "management plans". The resulting plans are rarely implemented. Although plans may incorporate all available technical information, that information is rarely developed or presented in ways that are accessible to water users and non-specialist decision-making audiences. This is important, because water users and non-specialist decision makers are generally the ones having de facto control over plan adoption and implementation. Unless these audiences both understand and have substantial ownership of plan proposals, development of a broad base of support for implementation will be difficult. Data and analysis need, therefore, to be presented in ways that are accessible to the array of potential stakeholders involved in any management context—not just towards a narrow base of technical specialists.

The central point here is that data and predictive analyses of hydrogeologic system behaviour play essential *social* as well as *technical* roles in planning and management processes. As Baker and Romm (1990) suggest, predictive analyses should be viewed as "negotiating texts"—agreed understandings of how groundwater systems work and what will happen in response to specific development or management interventions. Confident determination of sustainable groundwater extraction rates, pollution sources and rates of pollutant migration are essential for groups of different stakeholders to reach agreement. This requires solid conceptual understanding of the groundwater systems, detailed monitoring of hydrogeological paramters, and predictive analysis at aquifer system scale. Basic hydrogeological research, data and analysis are thus an essential foundation without which the social and political processes necessary to identify and develop viable management solutions cannot occur.

In addition to describing groundwater systems as a basis for management negotiations, analytical efforts need to focus on opportunities for management that could meet the needs of different stakeholder groups. The examples from California and Tunisia in boxes 23 and 24 illustrate this. In both situations, integrated approaches to water resource modelling have enabled the identification of management options that meet the needs of environmental as well as other water users. This type of analysis has the potential to greatly reduce conflict. By identifying innovative management avenues beneficial to a broad spectrum of user interests rather than simply identifying problems and their proximate causes, decision makers may be able to overcome the social and political obstacles to management.

6 Guidelines for addressing groundwater depletion and degradation

In order to develop effective approaches and institutions for groundwater management, the underlying factors causing specific problems and limiting society's ability to respond to them have to be addressed. Many of the proximate causes—expansion in well numbers, uncontrolled pumping, unregulated disposal of pollutants, etc.—can be addressed only through approaches that address more deeply rooted factors, including the common-property nature of the resource base, policy failures, perceptions, public awareness, variability, management capacity and adequacy of information. It is also evident that different levels of intervention are required for different levels of groundwater vulnerability. In cases where the resource base is highly vulnerable, immediate action to address proximate causes of depletion and degradation is essential. While some solutions may ultimately prove impractical, it is important to take whatever actions are possible while the resource base can be protected. These immediate actions will provide the time for longer-term solutions that address root causes to be developed and implemented.

Basic principles central to any attempt to initiate groundwater resource management are discussed here. Each principle does not belong to a checklist, but addresses a core requirement for management related to the following questions:

1. What is the **structure or context** governing management attempts and approaches?
2. **Who** needs to be involved?
3. **Why** is management important?
4. **How** are actions going to address real problems?
5. **When** are actions going to take place?

Figure 8 illustrates where each principle fits in relation to these basic considerations.

6.1 Structuring an approach to groundwater management

The need to integrate physiographic and socio-economic factors was highlighted in section 1.3. It was emphasized that this integration or "meshing" is not an academic exercise, but that the resultant framework can be developed into a progressive instrument for addressing critical needs in groundwater management. The design of workable approaches will invariably involve judgements on the inclusion and relative weight given to the principles listed below in the context of location-

Figure 8 Questions central to addressing groundwater depletion and degradation

Questions				
Structure What is the structure guiding approaches to the development of a management system?	**Who** Who needs to be involved in the process?	**Why** Why might they be willing to contribute to management efforts?	**How** How might individuals, organizations or sectors of society contribute to groundwater solutions?	**When** When can one expect effective solutions to emerge?
Answers (or principles)				
Strategic: An approach that is strategic rather than comprehensive.	**Stakeholders:** Local stakeholders must be involved because they often have the most direct control over groundwater conditions.	**Value and economic incentives:** Users respond to economic factors.	**Through a process of dialogue:** Unless dialogue occurs, the perspectives of key stakeholders are unlikely to be reflected.	**Over time:** Most management needs cannot be addressed rapidly.
Integration: Management approaches need to be integrated in a manner that reflects the interaction between hydrologic, economic, social and other systems.	**Policy makers:** Policy makers are often important because the incentives users face are shaped by policy decisions.	**Scientific information and data:** These are essential to understand problems. Without understanding people are often unwilling to act.	**Through implementation activities:** Involving a broad base of "stakeholders" in management implementation creates social ownership for it and brings critical resources.	
Flexibility is essential for management.	**Private-sector organizations** need to be involved because, as with stakeholders, they are often major users. In addition, they can often bring major resources to management initiatives.	**Education and social ethics:** Broad-based education is often central to building social support for management. In addition, individuals often act on the basis of social ethics or norms.		
Scale of the resource base and scale of the management response.	**Government and public-sector organizations** are central to management and need to be involved.			
Access to and responsible engagement with the resource base are essential. Understanding legal frameworks, rights and equity issues is essential for this.				

specific physical and socio-economic circumstances and the institutional responses they invoke (Livingstone, 1993). In this sense, the selected approach, the principles that underpin it and the processes it seeks to influence have to be measured carefully against available policy choices and management options in each setting. This is not as straightforward as, for example, river basin management, since boundary conditions, three-dimensional structure and temporal responses are much more complex for aquifer systems. This having been said, the principles outlined below provide a basis for starting to address groundwater management with some transparency and accountability, rather than putting the matter off until the resource base is seriously degraded. Applying the principles requires establishment of a strategic framework in which the links between all the principles and processes are clearly articulated. Such a framework transcends the physical and institutional frameworks and the processes that bind them and is designed to be *strategic* in addressing spatial, temporal and cross-sectoral complexity. It should be stressed that a strategic framework is much more than a logical framework that might apply, for instance, to the implementation of a sectoral programme or project. The strategic framework approach is elaborated below.

6.1.1 Strategic frameworks

Given the relatively short space of time over which aquifer systems have been depleted and degraded and the complex array of factors influencing aquifer conditions, management interventions will need to be strategic, as opposed to comprehensive. A strategic framework is needed to specify the intended relationship between interventions or management approaches and management goals. It should also specify the relationship between diverse sets of interventions—how together they are intended to form an approach to management that is internally consistent. It is the abstraction of key factors and targets (the who, why, how and when of management) and is intended to ensure that tensions and contradictions are identified and addressed. The fact that links (whether assumed or de facto) between objectives, interventions and intended outcomes are rarely spelled out constrains the development of groundwater management systems.

Figure 9 illustrates how the descriptive frameworks elaborated in the text (physiographic/socio-economic/legal/institutional etc.) relate to a strategic or conceptual framework. The descriptive frameworks represent a set of initial conditions that relate to broad social goals and are modified through an iterative process of pressure and intervention by different interest groups. The figure outlines the process of social and institutional change a strategic framework is intended to respond to and shape. The underlying concept is that divergences between conditions and social objectives (i.e., problems) eventually catalyse action by interest groups. This takes the form of pressure for policy, legal, economic and social reform. The outcome of reform initiatives, in turn, affects conditions (in this case, the status of groundwater resources). This then creates the "initial conditions" shaping subsequent iterations of the same process of change. This process has to be structured if iterations are to gradually close the gap between the context and social objectives (i.e., if management needs are to be addressed). As a result, development of a "strategic framework" is an essential precursor to effective groundwater management. In most cases, effective management will depend on a wide array of often distantly connected interventions—from education to law and technology to land-use planning. These interventions will need to be applied in

Figure 9 Descripitive model change process in groundwater management

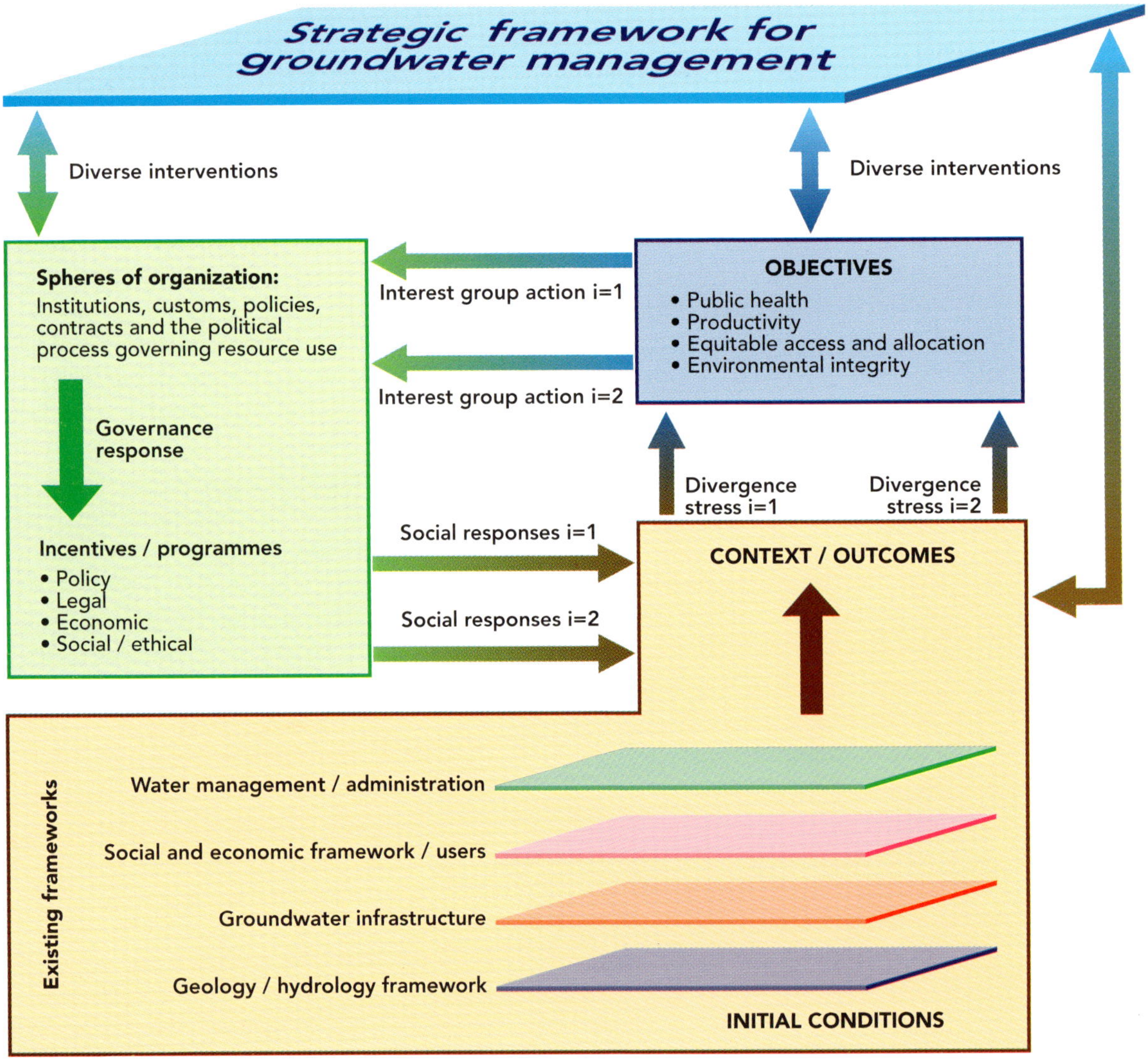

an equally diverse array of local contexts. Development of approaches that are internally consistent requires an underlying logical structure, a "strategic framework" to guide the selection of interventions in response to diverse conditions. Without some form of strategic framework, the logic underlying management interventions can be rapidly lost as the process of social change proceeds. In addition, policies and interventions often work at cross purposes. Identification of these contradictions is central to effective management.

To be effective, a strategic framework for groundwater management has to identify clearly: (1) the objectives that management interventions or approaches are designed to address; (2) the way diverse actions fit into the overall management strategy; and (3) the criteria against which the success or failure of specific strategies or interventions can be evaluated. In addition, it has to identify the mechanisms and processes through which such key issues as water allocation are to be addressed. A strategic framework provides the underlying logical structure necessary to develop action programmes or implementation plans. It serves as a reference point that identifies the logical basis for strategy components and enables re-evaluation. Development of one is not inherently a time-consuming or data-intensive process. A simple structure clarifying the connection that is assumed or exists between sets of actions and objectives is sufficient at the start. The framework should, however, be a "living" or iterated document that can be updated, refined and if necessary changed as information and experience are gained.

Strategic frameworks are implicit in most approaches and processes designed to address natural resource management needs. The complexity of groundwater management makes it particularly important, however, to identify explicitly the manner in which diverse sets of interventions are intended to address the even more diverse array of social, economic, institutional and hydrogeological conditions underlying both problems and their solution. Attempts to address complex groundwater management issues using such a framework have been initiated in India (World Bank, 1998a).

6.1.2 Integration and systemic perspectives

Integrated or comprehensive approaches are increasingly advocated as a basic principle for effective water management throughout the world (United Nations, 1977, 1992, 1998; World Bank, 1993). In most cases, however, integration focuses on the water-supply-and-use system itself—not the economic and social factors determining water demand or the distribution of pollution. While development of integrated approaches to the physical system is an important first step, it is equally important to recognize and integrate socio-economic and other factors that directly and indirectly affect surface-water and groundwater conditions.

In practice, development of fully integrated or comprehensive approaches to management is problematic (Mitchell, 1990). Basic data on hydrologic systems or water use are often unavailable. Social factors affecting water use tend to be complex and operate at multiple levels. The economic factors affecting water use, for example, are often a product of local features (transport and marketing systems), national policies (such as subsidy programmes) and global conditions (world grain prices). Similarly, the institutional context in which water management must be implemented generally contains multiple organizations with different mandates working at local, regional and national levels. This leads to fragmentation of responsibilities and often to conflicting interests.

Box 25 What is an integrated strategic framework?

An integrated strategic framework outlines the current understanding of resource and social system parameters and the interactions between them in a way that enables the logical structure linking problems, causes and potential interventions to be transparently understood.

A strategic framework needs to structure:
The *spatial* and *temporal* characteristics of:

- The natural resource base (groundwater and surface-water systems)
- Water-use systems (seasonal and other short-term or long-term changes in water use)
- Environmental systems
- Economic systems (the human activities dependent on water resources)
- Institutional systems (jurisdictional boundaries, legal structures and processes etc.)
- Social systems (urban and rural features, cultural groupings etc.)
- Emerging groundwater problems (overdraught, pollution, waterlogging etc.)

The *links* between the above characteristics and potential:

- Economic interventions (changes in prices, taxes etc.)
- Regulatory interventions (licensing, tech-nology choice, access limitations, allocation decisions)
- Institutional interventions (water rights, water markets, community and other management institutions)
- Technical interventions (water management structures, changes in end-use technologies)
- Knowledge interventions (educational programmes, attempts to develop conservation ethics etc.)
- Process interventions (participation, stakeholder involvement etc.)

The societal *objectives* to be addressed through water management and how resulting *allocation* patterns should meet:

- Basic human needs
- Social equity in resource access
- Environmental requirements
- Economic development

An integrated strategic framework needs to provide:
A strategy for initiating and guiding the iterative process through which emerging problems can be addressed and societal objectives met given the interlinked sets of systems. The strategy needs to identify where to start, what to start with, how to start, and how to continue once the process is started.

This complexity has fundamental implications for attempts to develop integrated approaches to water management and groundwater management in particular. In many cases, integrated water management initiatives degenerate into massive data collection and planning efforts that are out of date before they are completed. As a result, the need for integration must be balanced against practical limitations and the importance of immediate action to address specific problems. In sum, systemic perspectives and adaptive management approaches will be more effective than a "fully integrated" or "comprehensive" approach. A systemic perspective reflects the importance of interactions between hydrological and hydrogeological systems, economic systems and institutional systems in shaping water management needs and opportunities. In combination with adaptive approaches to management (approaches that are intended to be modified as conditions change or new information and insights become available), it can enable "integrated" approaches to management to be initiated relatively rapidly. It does, however, require a strong conceptual understanding of the broad array of physical, social, economic and institutional factors affecting water management needs and options (see box 25).

This consideration is very demanding on the institutions charged with making groundwater management effective. Such institutions require strong and stable institutional arrangements along with the necessary legal authority to make and enforce policy. They need to be knowledge-driven with broad access to data and information. They need to be capable of building widespread community support for courses of action across sectors and administrative jurisdictions. Finally, they require personnel capable of developing a broad interdisciplinary understanding of water management issues.

This last aspect is, perhaps, the most challenging. Throughout the world, most educational systems produce engineers, economists, sociologists and specialists in other disciplines. They produce few individuals with the broad interdisciplinary skills essential for developing integrated water management approaches.

6.1.3 Flexibility

A second major challenge is the need for flexibility. Because social, economic and hydrologic systems are dynamic rather than static and the factors directly or indirectly affecting groundwater conditions vary greatly among regions, management approaches must be flexible. Rapid changes in economic and social factors, along with recognized limitations on data availability and scientific understanding of management needs, indicate that management approaches will need to change as conditions and understanding evolve. The flexibility to change or modify approaches is, as a result, essential. Flexibility should thus be a core principle in any strategy for initiating action to address emerging groundwater problems.

Flexibility implies the development of policy frameworks that enable management approaches to be tailored to reflect the social, economic and physical resource conditions prevailing in different management areas (Moench, 1994a). National frameworks that attempt to specify management details—such as well spacing, specific prices for water, etc—will often mandate approaches that are inappropriate or unworkable at the local level. In contrast, policy frameworks that focus on broad principles and provide administrative or legal mechanisms enabling local managers to work out details are likely to be more workable and efficient.

A second dimension of flexibility has to do with the willingness of specialists and decision makers involved in groundwater management initiatives to accept participatory planning and to respect the perspectives and knowledge of all stakeholders—particularly local users. A balancing act is often required. In some cases, the perspectives of stakeholders are highly compartmentalized and self-serving (Kemper, 1996). There have to be fundamental checks on beneficial use, the public interest and the natural limitations of the aquifer systems to ensure that special interests do not dominate. At the same time, stakeholders often bring key insights to the process of developing management approaches. Furthermore, opposition by stakeholders (including the passive opposition of individual users) can undermine management approaches, whatever their technical merit. In many cases, approaches developed primarily on the basis of technical considerations lead towards a narrow array of such management approaches as centralized regulation. By emphasizing a limited set of responses, opportunities for initiating more effective management strategies are often missed. As some of the boxes in this section illustrate, in some cases it has proved possible to address groundwater problems through education, management techniques developed by local populations, changes in regional economic structures and other non-traditional mechanisms. In many cases, non-traditional mechanisms that rely essentially on communication and dialogue may be more politically, socially or economically acceptable than standard management techniques and can be equally viable responses from a technical perspective. Similarly, the viability of different institutional designs is often heavily dependent on cultural conditions. Management districts with locally elected boards of directors, for example, are common in the United States but might not work well in regions where they run counter to traditional systems of local governance. Overall, approaches must be flexibly adapted to respond to the opportunities and constraints inherent in local contexts.

6.1.4 Devolution to hydrologically viable scales

In dealing with most common-property resources, the complexity of management tends to increase with scale. This is a factor of group size and condition variability. Locating management activities at the smallest scale and lowest administrative level at which they can effectively be carried out is therefore key. This principle needs, however, to be balanced against the institutional importance of well-established resource boundaries which reflect the physical scale at which groundwater systems function. Clear boundaries for natural resource management units are as important for the development of effective common-property management institutions as they are for scientific understanding (BOSTID, 1986).

The scale issue has both technical and social dimensions. On the technical side, there is increasing recognition that groundwater needs to be managed at the scale of the aquifer. Defining aquifer boundaries is, however, a difficult and often subjective exercise. Large aquifer systems—the Gangetic basin in India or the Central Valley in California—can often be subdivided into much smaller management units. In many ways, the scale at which management activities should be organized depends on the specific management needs. Where pollution is concerned, for example, point and concentrated source problems can often be addressed at the local level within specific aquifer units. Similarly, it is often logical to manage cones of depression related to municipal pumping or sewage and industrial contamination of groundwater under urban areas at the level of municipalities, rather than at the scale of the aquifer. This having been said, it should be clear that local management initiatives need to reflect aquifer dynamics and fit within a management framework that recognizes the aquifer as the primary unit for management of the resource base. Ideally, for large aquifer systems, a single institutional entity (the equivalent of basin commissions for surface water) would have responsibility for developing the overall aquifer management framework and ensuring that local approaches are consistent. The absence of such an entity should not, however, reduce efforts to manage more localized problems at more localized scales, but the degree to which local initiatives can translate into positive impacts at the system level is not always predictable (Seckler, 1996).

Defining appropriate scales for management in cases of regional groundwater overdraught and/orsaline intrusion is particularly complex. In a broad sense, overdraught may be defined with respect to the water balance within a given aquifer, but the difficulties with this definition have already been pointed out (Miles and Chambert, 1995). Because the aquifer represents a single hydrodynamic system, over-abstraction cannot be addressed at management scales below the full system. This general statement, however, often masks real opportunities at local levels. In aquifers with relatively low hydraulic conductivities, lateral flow rates are often sufficiently slow that localized water-level changes have little impact on water levels in other parts of the same system—at least over the short to medium term. In this case, management interventions at local levels (below aquifer scale) are appropriate.

Part of the Eritrean rift in which localized groundwater occurrences sustain rural productivity

In sum, the physical scale at which management is needed depends greatly on both the physical understanding of the aquifer system and the specific problem to be addressed. After adequate studies, it may often be possible to address local water-level declines, salinity, waterlogging and pollution concerns through relatively localized sets of monitoring and management interventions. On the other hand, regional over-abstraction problems or salinity intrusion in coastal areas generally require co-ordinated actions at the aquifer scale. Identifying the principal characteristics of an aquifer source and the core issue that needs to be addressed—e.g., the specific water services (including environmental services) that are threatened and the types of management actions needed to protect them—is a key step in determining the appropriate response scale. In some cases, for example, regional water-level declines may not represent any real threat to the long-term sustainability of groundwater resources but may have major localized impacts on wetlands. In this case, it may be more efficient to focus on the development of sustainable water supplies for the wetlands rather than attempting to manage regional water-levels.

Hydrogeologic scale and the complexity of clearly defining boundaries interact closely with issues in the social scale of management. Ostrom (1993, p. 1908): "So long as the boundaries of who has rights to the water remain uncertain, no one knows what they are managing, or for whom." She goes on to note that "the presence of boundaries concerning who is allowed to appropriate from a resource has been used since the work of Ciriacy-Wantrup and Bishop (1975) as the single defining characteristic of 'common-property' institutions as contrasted to 'open access' institutions." In the groundwater case, the boundary issue is particularly complex. Conceptually, aquifer boundaries generally define the boundaries of the group with access to the resource—i.e., those who own wells on overlying lands. *Aquifer boundaries, however, rarely coincide with the boundaries of social and administrative units.* Again, the development of effective responses to problems may depend on pragmatic balancing of the scale of social and hydrogeologic units. Management units need to be socially as well as technically determined.

The complexity of organizing management institutions is generally inversely proportional to the size and number of stakeholders who are involved. Cultural considerations can also be important factors. For example, the degree to which different groups within society are accustomed to cooperation and the presence or absence of regional organizations or administrative units capable of assisting in management will determine specific management responses. In general, from a social perspective, smaller management units are likely to be more viable, particularly where they coincide with the boundaries of other pre-existing social units (village groups, municipalities, districts etc.). The social desirability of relatively small-scale management units must, however, be counterbalanced against the physical viability of management interventions at different scales. In most cases, management action needs to occur at intermediate or regional scales. Management units defined at the level of individual communities or villages will generally be too small from a hydrogeologic perspective and offer no economies of scale. At the other extreme, management of large regional aquifers as a unit may be neither essential from a hydrogeologic perspective nor practical in social and administrative terms.

6.1.5 Equity

The fundamental importance of water for basic needs was stressed at both the 1992 Earth Summit in Rio de Janeiro and the earlier United Nations Water Conference (Mar del Plata, Argentina, 1977). As recently as 1998, the Commission on Sustainable Development (United Nations, 1998c) has reiterated the point. Some analysts now place access to water at the level of basic human rights. Gleick, for example, recommends that "50 litres per person per day of clean water should now be considered a fundamental human right" (Gleick, 1995, p. 227). Invoking such "rights" may be arguable, but whatever the precise level of water reserved for specific uses, this position resonates with traditions that are deeply entrenched in many human ethical systems. As box 5 (above) on competing philosophies documents, both Roman common-law and Islamic tradition view access to water as a fundamental inalienable right. As a result, consideration of the equity implications inherent in management approaches should be a basic principle.

The distribution of access to land, soil and groundwater in Yemen is reflected in this Yemeni landscape

The importance of water for basic needs will always be a major consideration for groundwater development and management. But while the importance of equity is clear, effectively incorporating equity considerations into management strategies can be complex. Wherever possible, equity considerations should be given a strong foundation in water rights formulations and law. At the same time, it is essential to recognize that the establishment or reform of groundwater rights systems or groundwater law may not be feasible in many situations. At the very least, clear statements of equity as a core objective need to be incorporated within the strategic framework used to guide development of management systems. Measures of equity should be key criteria—along with sustainability and practicality—against which management proposals need to be evaluated.

6.1.6 Legal frameworks

The establishment of legal frameworks for groundwater management at the state and national levels should be a core principle in structuring groundwater management initiatives. In some areas, substantial changes in groundwater law are already beginning to occur. These include, for example, the development of permit systems, the creation of groundwater management districts and establishment of an array of fee systems for groundwater use. Change and the evolution of new legal systems are, however, laborious. Since new legislation usually creates friction with existing users (and eventually with prospective new users), the process of legal reform is delicate and often faces political resistance. Therefore, a common tactic in implementing new legislation is to respect, to the extent possible, existing uses, and to adjust them only gradually.

From reviews of current water law systems (Solanes and Villareal, 1999), legislation for the protection of groundwater resources generally contains the following key components:

- It establishes public ownership of the resource or declares a public interest in appropriate and sustainable management of the resource base. This is essential in order to create the legal standing for public or governmental entities to manage the resource in ways that reflect public goods, externalities and common-property characteristics inherently associated with groundwater.
- It contains recognition of private use rights. This is needed to give individuals the security essential for them to make investments and efficient use decisions.
- It links groundwater and surface-water rights and the administrative systems for managing them. This is increasingly recognized as essential because groundwater and surface-water are generally part of an interconnected hydrologic system.
- It establishes the conditions necessary to obtain and maintain a water-use right. This often includes a permit system reflecting such considerations as water availability (including natural variability), the beneficial nature of the use, the rights of other users, the payment of fees and tariffs, the duty to provide information, duties to respect instructions and orders from government or management districts, licensing of drillers, creation of protection, restricted and management zones etc.
- It contains provisions for quality protection. This often addresses waste and effluent disposal and may include provisions, such as sealing of unused wells, intended to limit migration of low-quality water.

The institutional structures created through legislation are much more variable than the above common components. Many countries attempt to impose uniform regulatory frameworks for groundwater management. These can be relatively rigid and impractical to implement under field conditions. In contrast, local water management institutions such as "districts" in the United States (Moench, 1994) and *Wasserverbande* in Germany (Barraqué, 1997b) can provide recognized frameworks through which a wide variety of management initiatives—from regulation to education—can occur. Legislation enabling the formation of groundwater districts in the state of Kansas (United States) is described in box 26 as an example of such enabling frameworks. Legal frameworks need to be designed in ways that accommodate natural variability and have sufficient flexibility to enable the evolution of different approaches in response to local conditions. Provisions 9-13 in the enabl-

Box 26 District enabling legislation: the case of Kansas

The Kansas Groundwater Management Districts Act is the enabling legislation for all groundwater management districts in Kansas. This act stipulates the process required to form a groundwater management district (GMD), funding and operational authorities and both specific and general direction for all activities either required or eligible to be undertaken. It is important to note that this act makes it a policy of Kansas that local landowners and water users should be allowed to determine their own destiny in regard to groundwater management issues as long as they do so from within a legally formed and operated GMD.

The basic requirement to form a local GMD is the existence of an aquifer system of sufficient size to support a district which is experiencing groundwater problems of a quantity or quality nature. If an area of the state demonstrates such a viable hydrologic community of interest, a local GMD can be formed. To date, five districts have been formed. The GMDs are operated under the direction of a locally elected board of directors from within the district. The only requirement for being a member of the board is that he or she must be an eligible voter as defined by the act.

In order to operate, Kansas GMDs must adopt a management programme, set an operational budget and then collect its fees. It is important to know that the management programme is locally written and then reviewed and approved by the Division of Water Resources, State Board of Agriculture, which checks local policy for consistency with the Kansas Water Appropriation Act. In this way, the local GMD programmes are coordinated with the state agency regarding groundwater management. To fund the districts, Kansas GMDs can respectively levy land charges not to exceed $0.05 per acre of land and assess water-use charges not to exceed $0.60 per acre-foot of water. Except for outside grants, gifts or contracts obtained or negotiated, the Kansas GMDs are entirely funded by local revenues.

The districts are granted authority in the following areas: (1) suing and being sued; (2) maintaining, equipping and staffing an office; (3) holding and selling certain property and water rights; (4) constructing and operating works for drainage, storage, distribution or importation of water; (5) levying water-use charges and land assessments, issuing bonds and incurring indebtedness; (6) contracting with persons, firms, associations or agencies of the state or federal government or private entities; (7) extending or reducing district boundaries; (8) conducting research and demonstration projects; (9) requiring installation and reading of meters or gauges; (10) providing assistance in the management of drainage, storage, recharge, surface-water and other problems; (11) adopting, amending and enforcing by suitable action policies relating to the conservation and management of groundwater; (12) recommending to the chief engineer (a state official) rules and regulations necessary to implement and enforce board policies; (13) entering upon private property for inspection purposes to determine conformance with policies; (14) seeking and accepting grants or other financial assistance from federal, public or private sources; and (15) recommending to the chief engineer the initiation of proceedings to establish an intensive groundwater-use control area. Under the act, the chief engineer may also, as a result of his own investigations, initiate formation of an intensive groundwater control area. Once established, the state has full regulatory authority in these areas.

Source: Note prepared by the Kansas Groundwater Management Districts Association, 1991.

ing legislation for groundwater management in Kansas (box 26) provide an example of how this can be done. As a basic principle, broad legal frameworks that enable management—such as legislation enabling formation of management organizations—should be established as early as possible. It may only be possible to develop more detailed frameworks after management experience has been gained from the original framework.

Enabling legal frameworks are particularly important where management is being attempted through organizations composed of local users. As Ostrom (1993, p. 1910) comments in the case of water-user groups for irrigation management: "Many water-user groups organize in a de facto manner but are not recognized by national governments as legitimate forms of organization. An effective irrigator organization lacking formal recognition may crumble rapidly when its authority to make legitimate rules for its own members is challenged and not supported by the formal government of a regime." In the previously noted Kansas case, locally managed districts have the legal authority under state water laws to adopt a wide range of regulatory powers. The process of forming management districts is initiated by 15 voters filing a declaration of intent to form a management district with the chief engineer of the state. Details of the proposal are then subject to review and minor modification by the chief engineer and the secretary of state. Following their approval, the proposal is publicized within the proposed district area and an election is organized. If it is approved by a majority of those voting in the election, the district is incorporated by the secretary of state and can begin operation as a legal entity governed by a locally elected board of directors.[16]

The approach outlined above evolved in a situation where landholdings are large, the number of actors who must be involved in district operations is relatively small and the formal structures for democratic processes are well established. As a result, it may be difficult to implement or inappropriate in many contexts. An example of an alternative approach that balances both state and local interests is presented in figure 10. This approach is designed to provide local stakeholders with an officially recognized process for forming an organization and developing a management plan. This plan would be subject to technical approval by a state organization. If local users cannot agree on a technically viable management approach, the approach enables the state to initiate management action. The approach also contains an independent body for appeals. This component—and its independence—is essential in order to minimize disputes between local communities and government organizations.

The above approach reflects a pragmatic response to balance of power and governance considerations that directly affect the success of attempts to manage groundwater resources. Governments often lack the political or police power to control groundwater use and abstraction directly. Wells are widely dispersed and generally on private lands. Whatever the political philosophy of any given government, attempts to directly control use are frequently impossible to implement or enforce. Yet, stakeholders' participation is no panacea. Lacking information on the complicated dynamic nature of groundwater systems, stakeholders are often on unstable scientific ground. Furthermore, while users may implement manage-

[16] Kansas Groundwater Management Districts Act (KSA 82a-1020 et seq.).

Figure 10 Institution formation approach example

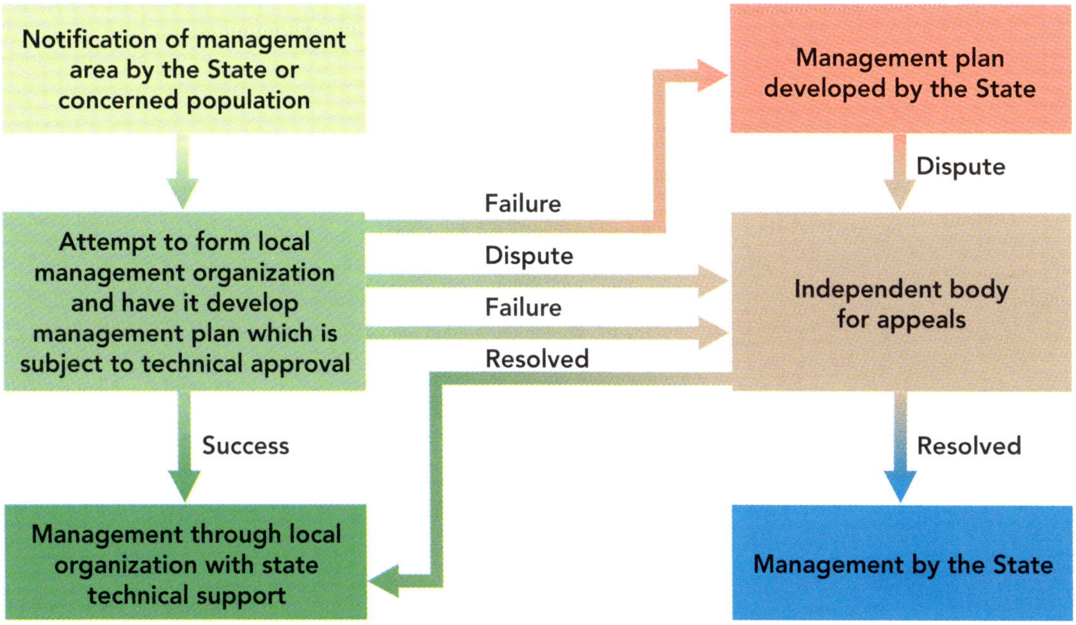

ment measures, they often will not do so if inappropriate financial incentives, such as irrigation subsidies, promote the unrestricted exploitation of water resources. In addition, they will not be able to implement management measures if there is no public legislation backing their activities or if they lack some sort of publicly recognized organizational structure. Private associative arrangements, such as the water-user associations often being formed in surface irrigation systems, are inadequate for this purpose. They lack the legal means to compel appropriate behaviour by users reluctant to behave according to the needs of sustainable management. They also lack the legal capacity to compel financial contributions from users not willing to pay. That is why successful stakeholder participation has usually been clothed under public law, backed by adequate information (usually developed by government administrations) and supported by adequate water legislation.

6.1.7 Water rights

Evaluation of water rights systems and reform options should be a central principle in the process of developing groundwater management approaches. Legal frameworks for management are closely connected with water rights issues. In virtually all situations, water rights systems exist. The rights system can be implicit (a function of cultural norms or regulations and permits) or it can be explicit (encoded in law at any level from a national constitution down). Most water rights systems evolved prior to the emergence of management needs and poorly reflect basic complications stemming from the common-property nature of groundwater resources. As a result, rights reform is often essential for effective groundwater management.

In most parts of the world until the beginning of the twentieth century, groundwater was considered either as an appurtenance to surface land rights or as a right of the driller of a well. Regulation and control were not a priority and in some jurisdictions, such as Mendoza, in Argentina, statutes for groundwater protection were deemed unconstitutional since they interfered with private property. Only exceptionally were private rights restricted, on grounds of malicious abuse, intent to damage, or, in the specific case of Muslim law, humanitarian concerns, such as the right to thirst.

Civil code system countries, such as Continental Europe and South America, have generally allocated groundwater domain on the basis of landownership. Thus in Argentina, France, Italy and Spain, groundwater rights follow ownership of the overlying land. The above approaches can be traced to common-law (originally Roman) sources. Under these traditions, no separate rights were recognized to "diffused percolating groundwater", since such groundwater was considered a mere ingredient of the soil. Therefore, the owner of the surface estate was also the owner of the water. Consequently, damage to third parties, other interests or other users caused while using groundwater was largely irrelevant (Murphy Earl, 1991). This issue has emerged as a basic flaw in many rights systems as third-party impacts have increased with development.

The role of water rights as a foundation for the development of formal water markets is also important to recognize. Without a clear water rights system, the right of individuals or regions to transfer water from existing low-value uses to new, higher-value uses is often unclear. Since water transfers between users and sectors will be increasingly essential over coming decades, transparent water rights systems will be of increasing importance.

Although the presence of flaws in existing rights systems and reform needs are clear, the practical difficulties posed by instituting reforms should not be underestimated. Part of the problem is inherent in the nature of groundwater resources. In most cases, aquifer dynamics are poorly understood. Groundwater is a flow resource. As a result, changes in use by any single user or group of users have third-party impacts, the magnitude of which depend on the flow conditions in the aquifer systems. Crafting rights systems in ways that adequately reflect these third-party impacts would be complex even if resource dynamics were fully understood. A second dimension of the water rights problem has to do with the large numbers of individual users present in such countries as India. In this case, the physical complexity and cost of monitoring water use may pose insurmountable obstacles to establishment of quantitative rights over groundwater. Finally, in many countries establishment of any private groundwater rights system would run counter to deeply rooted customs and ethical structures (see box 5, above).

Thus, it is highly important for legal frameworks to be established to enable rather than block the initiation of much-needed management by different groups. As with the establishment of detailed legal frameworks, more complex water rights issues may only be possible to address as management experience is gained. This said, it is equally important not to ignore fundamental water rights issues until existing water rights can be reformed or a new formal water rights structure can be developed. Equity issues are a case in point, and represent the final basic structural principle that needs to be reflected in any management attempt. This issue is dealt with in section 6.1.7.

Stakeholders are not always obvious. This Nepali fisherman's livelihood depends on in-field irrigation.

6.2 Who needs to be involved

In addition to the overall approach, the question of who needs to be involved in the development of management systems is fundamental. In a general sense, four groups stand out and should, as a basic principle, be involved in management initiatives:

- Local stakeholders—water users and others whose interests are directly affected by groundwater management and whose actions often determine the effectiveness of any given initiative;
- Policy makers—those who have the ability to influence the institutional environment within which management approaches must evolve;
- Public-sector organizations—these stakeholders often have their own internal agendas and control large programmes that either directly or indirectly have major impacts on water resources;
- Private-sector organizations—these stakeholders are often major water users whose interests may or may not coincide with those of local stakeholders.

The roles these groups can play in the development of systemic and adaptive groundwater management approaches are discussed on the next page.

6.2.1 Identification and involvement of stakeholders

Because groundwater is a common-property resource and the de facto power to control its condition effectively is dispersed among users, stakeholder involvement and education are essential for any attempt to manage the resource base. Stakeholder involvement should not, however, be interpreted as implying a need to involve every well owner, water user or polluter in the formulation of management initiatives. It is essential to educate and involve two types of stakeholder groups: (1) those whose actions individually or as a group have a major impact on the condition of the resource base; and (2) those whose interests will be significantly affected by management regimes. The above groups are not mutually exclusive, although they often have different characteristics. The first category generally consists of large water users, such as municipalities and large industrial and agricultural interests. They tend to be relatively well organized and have substantial social, economic and political power. The second category tends to be much more dispersed and, in many though not all cases, less socio-economically powerful. It includes diverse interests ranging from small farmers to peri-urban dwellers to environmental non-governmental organizations. Users who are significant polluters often are a major exception to the above pattern. In the pollution case, dispersed users can have a major impact on the overall condition of groundwater resources, such as in the case of farmers' application of agricultural fertilizers, herbicides and pesticides.

A key principle in attempting to identify and involve stakeholders is to start by being as *inclusive* as possible. Some stakeholders—large users or industrial polluters—may be relatively easy to identify and will tend to include themselves in any process they see as potentially affecting their interests. Educating this group is important, since they have the political and economic power to block the development of effective management strategies and are likely to do so unless they fully understand management needs. Proactive moves to identify, involve and educate the second category of stakeholders and the small individual polluters are equally important. Groundwater management initiatives will fail unless they influence the behaviour of stakeholders whose actions individually or as a group have a major impact on the condition of the resource base. At the same time, if management initiatives are controlled by large water users and polluters, they will tend to protect the interests of those users at the expense of less socio-economically powerful sections of society and environmental interests.

Beyond equity and political considerations, stakeholder involvement can often bring unexpected resources and insights to management. In Yemen, for example, local communities often manage water-supply systems and a few have implemented schemes for protecting the aquifers on which drinking-water sources depend from exploitation by agriculture (Moench, 1997). In Gujarat, India, there is a large farmer movement to recharge dug-wells in hard-rock areas (see box 27). More importantly, traditions of water management are widespread at local levels in such countries as India and can provide a basis for the development of more comprehensive approaches (Agarwal and Narain, 1997). Stakeholder involvement does not imply an absence of regulation. As Barraqué notes: "Today, community-based water management is increasingly advocated for its efficiency, and it has a legal base in most European countries. It supplements the move towards growing public water regulations" (Barraqué 1997b, p. 2).

Box 27 Saurashtra's well recharge movement

A major local initiative to address groundwater overdraught in Gujarat has emerged through a spontaneous and self-propagating people's movement for groundwater recharge that began in the Saurashtra and Kutch areas around 1990 and has since been spreading to northern Gujarat. The recharge movement is driven by deeply rooted ethical and religious concepts. It was catalysed by the Swadhyaya Parivar, a relatively young spiritual movement, and has been subsequently supported by many civil society institutions and Hindu religious organizations. The movement is based on traditional Hinduism, and embraces the concept of instrumental devotion, which imbues everyday actions undertaken for the common good—such as environmental protection—with religious merit. This apparently simple idea led to the creation and maintenance of numerous "common properties", including groundwater, in rural areas of Gujarat and Saurashtra, with voluntary labour, donated materials and funds.

As early as 1978, Swadhyayees promoted the slogan "If you quench the thirst of Mother Earth, she will quench yours." Swadhyayees found this slogan prophetic 10 years later when three successive drought years (1985-1987) brought water issues to their cyclical peak in the public mind. Taking a cue from Israel, Swadhyayees began to adapt and improvise on techniques used the world over for *in situ* rainwater harvesting and conservation. "The rain on your field stays in your field; the rain on your roof stays in your home; rain in your village stays in your village" was the slogan the Swadhyayees of Saurashtra adopted. Many Swadhyayees—mostly poorly educated farmers—began trying out alternative methods of capturing rainwater and using it for recharging existing large-diameter dug-wells. In the 1989 monsoon, there were isolated experiments throughout Saurashtra; but in some Swadhyayee villages, the entire community tried out such recharge experiments, and they found the results highly beneficial. As the word of such experiences spread, new villages began joining the movement. Many non-governmental organizations also came forward to help by bringing water issues to public attention, often in collaboration with the media. Non-governmental organizations also supported local "people's" innovations to augment *in situ* recharge of water, using wells, tanks, rivers, sumps, soak-pits, roofs and other mechanisms to capture water and divert it to open dug-wells for aquifer recharge. After initial years of chaotic propagation, the Swadhyaya Parivar itself devised a loose organization to systematically spread the message and the method.

There are no formal studies yet of the actual scale of well recharge work; however, many different sources suggest that between 1992 and 1996 over 90,000 wells were recharged in Saurashtra and some 300 *nirmal neer* (farm ponds for recharge) were constructed. Enthusiastic Swadhyayees set themselves a target of over 125,000 wells and over 1,000 farm ponds during 1997. It is widely believed that regularly recharging 500,000 wells in Saurashtra during the monsoon can solve the groundwater problem of the region. At the rate at which the movement is growing, it would not be surprising if that target is met before the turn of the millennium. By 1993-1994, the well recharge movement had begun spreading northward to other parts of Gujarat. At present, large amounts of anecdotal evidence highlighting the beneficial economic impact of recharge is spreading by word of mouth and through local-language media.

Ultimately, at the basin level, the number of stakeholders directly involved in the development of management initiatives will need to consist of a core group representing both major users and those affected by management decisions, including surface-water users. Research on common-property management institutions has found that the ability to build a consensus sufficient for effective management decreases as group size increases (BOSTID, 1986). Attempts to formulate strategies often fail if the involved group of stakeholders is too large for any degree of consensus regarding management to emerge. Inclusiveness and equity are of equal importance to success. A balancing act is often required in which efforts are made to ensure both broad-based involvement and the ability to generate agreement regarding management directions.

Who holds the "stakes" in this balancing act at this spring-box in Equateur Province, Democratic Republic of Congo?

6.2.2 Involvement of policy makers

Policy makers are a second major group whose involvement is critical. The development of groundwater management systems generally raises basic policy issues. These can include, for example, legal issues regarding the right of management organizations to come into existence or to regulate water use by well owners and governmental issues regarding the role of different public-sector organizations. Often, policy issues will emerge only in the course of management initiatives. As a result, continuous involvement of policy makers can be critical to success.

Specific mechanisms for ensuring policy-maker involvement depend heavily on the local governmental context. In some cases, as in the Joint Forest Management initiatives in India, formation of networks that include key policy makers and working groups to oversee pilot management initiatives can be a very effective mechanism (Poffenberger and McGean, 1994). Where it is more difficult to involve policy makers on a regular basis, workshops and publications directed specifically towards them can be useful. In any case, identifying the most effective mechanism to ensure awareness on the part of policy makers and their regular involvement in addressing the policy issues central to management should be a basic principle in management initiatives.

6.2.3 Encouragement of private-sector and public-sector collaboration

The common-property nature of the resource base and the public good characteristics of many of the services dependent on it imply that government, local authorities and other community and public institutions need to be involved in the development of management solutions. At the same time, many inputs essential for groundwater management, such as technologies and technical advice, can be provided by the private sector. Indeed, most groundwater users operate as part of the private sector. Their technological and operational choices can be heavily influenced by the presence or absence of other private-sector actors. Farmers, for example, often cannot adopt improved irrigation technologies or low-water-using crops unless associated servicing, processing and marketing industries are well established. Governments can play a key role in catalysing the establishment of these private-sector functions that enable individuals to change water-use patterns.

Some of the most effective initiatives to address groundwater over-abstraction in the United States have occurred as a result of public/private-sector partnerships. Research on effective local techniques for improving irrigation efficiency has, for example, been a key factor underlying the establishment of commercial enterprises for equipment supply. These enterprises were, in turn, essential in order for farmers to gain access to new technologies. In the High Plains Underground Water Conservation District area of Texas, education and technology shifts brought about by private/public-sector partnerships have arrested declines in the Ogallala aquifer under the district's management area (see box 26). The control of industrial pollution can also involve good partnership between the public regulatory bodies and the industrial sectors in developing solutions that minimize environmental impacts on aquifer systems and save processing costs through water conservation measures (Porter and van der Linde, 1995).

Recognition of the respective contributions of public and private organizations to the development of effective management initiatives is a basic principle in the development of groundwater management. Alone, neither private-sector nor public-sector initiatives will be sufficient.

6.3 Why: causes and incentives

The question of "why" groundwater problems have emerged or "why" water users might support or oppose different courses of action is often poorly addressed by those seeking to develop management initiatives. This is, however, central to the development of approaches that are socially and politically feasible.

Although numerous factors will influence the willingness of any individual to support management, awareness of the value of groundwater and understanding of its role and the basic nature of the resource appear particularly important. Groundwater is often undervalued (National Resource Council, 1997). In addition, both public understanding and scientific information on the resource base are often lacking. For these reasons, valuation, education and improvements in scientific information are key points of leverage to influence the reasons "why" different groups may support or oppose management.

6.3.1 Valuation and economic indicators

Quantifying the economic value of undervalued groundwater resources and identifying those instrumental values that are not possible to quantify should be a basic principle in groundwater management initiatives for several reasons. First, value estimates can play a major role in focusing policy-maker and public attention on threatened resources. Second, valuation often influences allocation and can be a key mechanism ensuring that water "flows" to high-priority uses (as opposed to merely "efficient" uses). Third, value estimates are critical in order to evaluate the level of investment in groundwater development, protection, monitoring and management that can be economically or socially justified. As the National Research Council Committee on Valuing Groundwater notes in its recent publication *Valuing Ground Water* (1997), in some cases pollution legislation (the "Superfund" in particular) in the United States has placed an essentially infinite value on groundwater resources. In other cases, the value of groundwater is virtually unrecognized. All factors being equal, management should be directed towards protecting those groundwater resources that have the highest social, economic or environmental value.

While valuation does not have to be either economic or quantitative, a systematic process for evaluating and prioritizing the relative importance of different aquifers requiring management is fundamental. Many quantitative techniques are available that can provide at least first-order estimates of the economic value of groundwater in different contexts. These techniques are summarized in *Valuing Ground Water* (National Research Council, 1997), but at best provide partial estimates of groundwater value. Dubourg argues that, at least in some instances, groundwater is a *critical* resource in that damage or depletion of the stock cannot be compensated by other forms of natural or man-made capital (Mähler, 1986; Dubourg, 1997). Where groundwater is a critical resource in this sense, techniques for estimating market and non-market use values may not reflect the "true" value of the resource. In addition, valuation techniques are inadequate to address the fact that many values or water services are products of groundwater system status and do not necessarily relate to volumetric patterns of water allocation. Base flows in rivers may, in certain cases, be closely related to adjacent groundwater flow patterns and only weakly related to overall volumes stored in an aquifer. A deep alluvial aquifer in a semi-arid region may, for example, contain a large stock of stored groundwater but contribute only a small amount to the base flow in surface streams. In contrast, a shallower alluvial aquifer with much less total storage may contribute far more to base flows if it recharges and drains rapidly on a seasonal basis. Valuation of the groundwater in storage based on its value for specific applications (such as irrigation) would, as a result, ignore the value associated with base flows.

Valuation issues are closely linked with the use of non-market economic incentives as key tools for groundwater management. From the perspective of individual users, subsidized pumping costs and regulated water charges (such as licence fees for access and/or amount used) are often the only direct indicators of groundwater value. Changing these costs is, as a result, a key mechanism for communicating the larger value of the resource to individual users. Unless this value is communicated to users directly through the prices they must pay for groundwater use, many will choose to ignore exhortations to manage their demand.

Economic incentives are a key factor driving the extent and efficiency of groundwater use. In some districts in the western United States, the price of water from surface canal systems is used to encourage sophisticated corporate farmers to shift between groundwater and surface-water sources as part of a management system. In India, a large debate has been taking place over the past five years concerning the role of energy prices and water markets in groundwater-use efficiency (Moench, 1995b; Moench, 1994b). In Germany, 11 Länder (regional governments) have set up abstraction charges on water-supply. In some, such as Baden-Württemberg, subsidies are used as incentives to control agricultural pollution. Denmark also uses abstraction charges (Barraqué, 1997b). Overall, economic incentives are increasingly recognized as a key mechanism for achieving groundwater management objectives.

Water markets are another mechanism for communicating the value of groundwater to users, but generally under very specific conditions. Detailed discussions of water markets beyond those contained above in section 5.1.1 and in box 21 on the differing roles of groundwater markets, are beyond the scope of this review. Two points are, however, important to emphasize. First, the evolution of large-scale formalized water markets depends on establishment of a clear system of transferable water rights. This is complicated and, as discussed in section 6.1.5 on water rights, faces numerous practical hurdles. Second, water markets will not, in themselves, resolve valuation and allocation problems. In most cases, the value of groundwater established through a water market will only reflect short-run values that benefit individual users. They will not reflect public values, such as environmental protection, or long-term intergenerational values, such as sustainable management of the resource base. As a result, water will tend to be allocated to uses that bring high short-run returns to individual users.

6.3.2 Public and policy-maker education

While the importance of educating stakeholders is acknowledged, it is equally important to educate and raise awareness of the public and policy makers as a basic principle for effective groundwater management. In most societies, only a small fraction of the population is aware of groundwater or the economic and environmental activities dependent on it. The actions of this broad population—from their day-to-day pattern of water use to their political support for management—

are of fundamental significance for the success of groundwater initiatives. As a result, educational approaches that address this broad group are necessary. But such approaches need to go beyond traditional audiences and mechanisms. It is as important for technical specialists to be aware of the social and economic dimensions of management as it is for policy makers or local communities to be aware of the array of technical options. In most cases, education needs to be a multidirectional dialogue in which policy makers, technical specialists, water users and other stakeholders come to understand the multiple dimensions of groundwater problems and management opportunities.

In some cases, education may be a core mechanism for management in itself, rather than just a part of the process in reaching management agreements. In the Texas High Plains, for example, governmental options for regulation and direct management are limited. There, innovative education initiatives implemented over the past 20 years have been a key factor enabling overdraught problems to be addressed.

The High Plains Underground Water Conservation District (HPUWCD) in northern Texas provides an interesting example of successful overdraught management based on education and the promotion of conservation technologies. This district lies over the Ogallala aquifer and has no access to major surface supplies. Furthermore, under Texas law all rights to groundwater are rights of capture based on landownership. Because of this, allocation and other strict controls on use are not management options. Lacking regulatory power, the district focuses on conservation. It has developed a two-pronged approach consisting of: (1) working with researchers to develop appropriate conservation technologies; and (2) the promotion of those technologies via education, economic and ethical approaches.

Staff of the HPUWCD work closely with both local research organizations and farmers to develop appropriate and low-cost conservation technologies. In addition, they have emphasized conservation ethics. The district does everything to create an awareness that wasting water is, to use the district manager's words, "going against mother, God and country". The district has a huge education programme in the schools and is also disseminating information through the media. It has also constructed and installed weirs in highly visible locations so that farmers can see all unutilized water running off their fields.

The district has little regulatory authority and less inclination to resort to direct policing of groundwater withdrawals. Despite the minimum regulatory approach, average well levels have declined at an average annual rate of only 0.49 feet over the past 10 years and not declined at all over the past 5 years (HPUWCD newsletter, May 1990). This is in distinct contrast to the annual declines of 3 to 4 feet which were commonplace in the 1960s (HPUWCD newsletter, November 1989).

Finally, it is important to recognize that the first step in educating society is educating children. In the United States, such organizations as the Groundwater Foundation, which actively promote children's education on groundwater topics, play a critical role in shaping the future ability of society to address groundwater management needs.[17]

---

[17] The Groundwater Foundation, P.O. Box 22558, Lincoln, Nebraska 68542-2558.

Furrow irrigation from groundwater with microprocessor-controlled surge flow to obtain uniform applications along the whole length of the row.

6.3.3 Targeted improvements in scientific understanding

Data on groundwater systems are in most cases far from adequate for management. At the same time, it is difficult to collect all the necessary data in advance or to identify the types of information that may be required. Baseline monitoring of key parameters is essential in order to signal the emergence of potential problems. Beyond this, data collection needs may emerge only when indicators signal potential problems or as management actions are initiated. Furthermore, even where the types of data potentially needed are clear, collection may be economically viable only when indicators signal specific needs. A combined strategy involving baseline monitoring and progressive refinement of information—targeted data collection and research—as needs emerge is thus a basic principle in any action strategy for groundwater management. These two sets of activities were discussed previously in section 5.7. Two elements of this strategy are, however, emphasized below.

*1. **Basic hydrological and hydrogeologic data collection and research are essential.*** This includes: long-term monitoring of groundwater (levels and quality) and surface water (flows and quality); the synthesis of background information (hydrogeologic and socio-economic data); and detailed investigations in areas where indicators signal the emergence of specific problems. Operational and basic research are also essential in order to document and evaluate the results of management attempts and in order to understand key elements of the hydrogeologic process that underpin various management options. The main factors that need to be included in any research and monitoring programme are outlined in box 28.

*2. **It is essential to monitor resource use and potential sources of pollution.*** Basic information on how groundwater resources are being utilized and who is using them is as essential for management as scientific data on the aquifer system itself. Without understanding how groundwater resources are being used, particularly for irrigated agriculture, it is impossible to identify points of management leverage. Devising management systems requires detailed knowledge of such factors as the actual locations where groundwater extraction is occurring, the efficiency with which it is used and its role in agriculture and other water-use systems. Programmes for registering wells and estimating—either directly or indirectly—the amount of groundwater extracted are essential first steps. Similarly, it is important to identify, as far as possible, other activities that could have substantial impacts on groundwater conditions. This includes the basic information on potential point and non-point sources of pollutants arising from industry, domestic sewage and agricultural chemical use.

Box 28 Know your resource

The ability to manage groundwater resources depends on solid, scientific conceptual, data and predictive foundations. Information essential for each of these includes the following.

Conceptual foundations

- Hydrogeologic maps identifying aquifers, the characteristics of geologic formations and major surface-water features
- Well logs identifying formation characteristics and water quality variations at depth
- Major water-use patterns—key environmental features, cities, agricultural areas and industries
- Potential points of contamination or migration of low-quality waters

Basic monitoring data

- Climatic parameters, including precipitation, evapotranspiration, sunlight hours and humidity
- Surface-water flows at key points in all major streams and conveyance systems (such as canals)
- Groundwater levels including seasonal fluctuations annual changes (the photo illustrates the collection of high temporal resolution of levels in wadi underflow to monitor "flashy" recharge events in Oman)
- Groundwater quality (electrical conductivity plus any additional parameters that the conceptual foundations suggest should be monitored)

Predictive analysis

- Water-planning models capable of integrated analysis of water demand, use and supply systems, such as the WEAP modelling system produced by the Stockholm Environment Institute
- Hydrologic models for detailed analysis of groundwater flow patterns, specific aquifer conditions and surface stream hydrology
- Ecological models for predicting the impacts of water resource management interventions on key environmental values

Beyond the data and analytical techniques themselves, the role monitoring and analytical systems can play depends heavily on their impartiality. Water management issues are often highly politically and socially charged. Agreements on management often depend on basic data and analyses that are scientifically objective. Strong, independent scientific organizations (as distinct from specific users such as agriculture or municipal water supply) are often essential to ensure that data and analyses are viewed as objective by stakeholders. The limitations of predictive analyses must also be recognized. Model results are often presented and viewed as "fact". The accuracy of models depends, however, on numerous factors, including the validity of conceptual foundations, data adequacy and programmer skill. As a result, the degree to which predictive modelling studies can be relied on needs to be carefully evaluated, based on the specific management questions of interest and the conceptual and data inputs required to answer them.

6.4 How: steps towards solutions

6.4.1 Process

Perhaps the most fundamental principle to recognize for starting practical groundwater management initiatives is the importance of long-term process. This recognition is reflected, for example, in the IRP processes developed by such organizations as the Metropolitan Water District of Southern California, where management involves a complex mix of groundwater and surface-water sources as well as a wide variety of social, political and economic considerations (Metropolitan Water District, 1995). Any long-term process will lead to different solutions in different areas and will generally reflect considerations specific to each of those areas. While many of the details of management solutions and institutions that emerge will differ, inputs required to enable the process to take place and many of the mechanisms through which management is implemented may be much more broadly replicable. Starting the process by initiating some form of management initiative is the essential first step. In addition, as discussed in section 7.1, some form of organization to guide the process is important. Formation of a nodal group or network that includes a variety of stakeholders, policy makers and practitioners at an appropriate scale can be an important starting point.

In addition to the process itself, ongoing efforts to build capacity will be essential in most regions. As previously noted, the capacities required for management often differ from those developed earlier for groundwater exploration. Groundwater departments developed to support exploration activities often lack many of the technical, administrative and social capacities essential for management. The need for capacity-building with regard to the socio-economic dimensions of management is likely to be particularly critical. In virtually all parts of the world, water management has been regarded as primarily a technical subject. While the technical dimensions remain important, implementation of management activities depends primarily on action by individual users. This is particularly true in the groundwater case, where use is controlled by individuals and demand-side management is contemplated. Concerted action by stakeholders is highly unlikely unless social and institutional capacities match technical capacities.

6.4.2 Implementation

The importance of process must be balanced against the need for implementation. The danger in emphasizing process is that it can be endless and never lead to actual activities on the ground. Implementation, despite its obvious importance, is often treated as an activity to be initiated only after studies have been completed and a "final" course of action has been decided. Implementation should, on the contrary, be viewed as an integral part of the process of developing effective management systems. Approaches need to be continuously tested and refined. This can be done only in the context of learning gained from practical experience with implementation of projects, policy development or other actions. Implementation activities are thus essential at all stages in attempts to address groundwater degradation. Starting implementation initiatives early in the process of developing a management system and using those early initiatives as a source of information on approaches and needs should, thus, be a basic management principle.

6.5 When: the length of the process and the role of indicators

Time is essential for institutional development, for implementation experiments and for aquifer systems to respond to management. Most sustainable solutions will evolve only on the basis of long-term initiatives designed to address the underlying causes of emerging problems. Years may be required. Recognition of this is a basic principle on which efforts to initiate management must be founded.

Timing is also important. The opportunity to take effective action often depends on the conjunction of major social events or political changes. The clear identification of management needs and potential actions to address them is essential in order to take advantage of the windows of opportunity as they emerge. Selected indicators play a central role in this.

Policy makers and society as a whole become aware of groundwater management needs only *after* problems have reached a point where they are difficult, and in some cases impossible, to address effectively. Indicators that can catalyse management interventions before this occurs are needed. For this reason, it is important to develop indicators "of source vulnerability". In a variety of hydrogeological settings, the effects of development on groundwater flow patterns and groundwater quality can be simulated or predicted on the basis of relatively limited data sets. Although precise understanding of the groundwater flow and degradation mechanisms is generally limited, there is sufficient experience to justify the development of vulnerability indicators. These can be based on a combination of basic hydrogeological information (boundary conditions, flow and migration parameters, water-level fluctuations and changes in water quality), water-use data (existing and projected water demands on sources for domestic, irrigation and other uses) and potential contaminant loads (discharge of waste water and the location of chemical-intensive activities). Some of these indicators could even be based on projected settlement or land-use patterns providing warning before activities even develop. Combinations of indicators can provide an integrated picture that ties in well with national and subnational water and land-use planning exercises. These types of indicators greatly enhance monitoring strategies and can be used to identify potential impacts on groundwater and associated environmental resources before they occur.

What are some basic warning indicators that can be monitored at the start of any initiative to manage groundwater and help to focus attention on specific areas before problems emerge? The following list is not exclusive but should provide a starting point.

- ***Rapid increases in well numbers.*** In most situations where well numbers are increasing rapidly, particularly for high-volume uses such as agriculture, substantial potential exists for mobilization of low-quality groundwater bodies and, over time, the development of overdraught conditions. A rapid increase in well numbers is therefore a key advance indicator that groundwater management will be needed.
- ***Land-use changes from low to high intensity.*** Land-use changes from low-intensity activities, such as subsistence agriculture, to higher-intensity uses, such as industry, municipal use and commercial agriculture, generally signal an increase in potential pollution sources and often also an increas in groundwater extraction. Whenever this occurs, the potential for groundwater pollution increases substantially. As a result, changes in land use, particularly where they can be projected in advance, are generally a good indicator that groundwater management needs should be evaluated.

- ***Changes in groundwater levels.*** Rising or falling trends in groundwater levels beyond normal seasonal fluctuations are a good indicator that management may be needed. Declining trends often signal the emergence of overdraught or the potential for mobilization of lower-quality and deeper groundwater and, where groundwater levels are relatively shallow, rising trends signal potential waterlogging. It is important to note that substantial water-level changes often emerge only after problematic use patterns are well established. As a result, while substantial changes in water levels are frequently a good indicator of the need for management, they give limited advance warning of potential problems.
- ***Water quality trends.*** Increases in contaminants such as total dissolved solids, bacterial contamination, nitrates etc. are clear indicators that management is needed. Frequent "spikes" in pathogens or other contaminants are also a good indicator of management needs, even where the overall quality of groundwater appears to be remaining high.
- ***Mobilization of low-quality water bodies.*** Saline intrusion in coastal zones, for example, is characterized by gradual landward movement of saline ocean water. The water quality of the coastal aquifer may remain stable in some parts of the coastal aquifer while this is occurring. Frequently, only changes in the location of the salinity front and the quality of water in wells underlain by the front signal that use patterns are disrupting the overall hydrologic balance. Equally, overdraught of shallow phreatic aquifers may induce lateral or vertical intrusion of lower-quality groundwater from adjacent or deeper formations.
- ***Significant changes in the hydrologic balance.*** Wherever water-balance calculations indicate that groundwater abstraction is a significant or major portion of the volume of renewable resources, management needs should be investigated. While water-balance estimates are often used as warnings of overdraught (e.g., where abstraction approaches or exceeds recharge estimates), it is important to recognize that mobilization of low-quality water may occur when extraction is a relatively small portion of recharge. It is also important to recognize that in most situations sufficient data to calculate the water balance in specific aquifers accurately are unavailable. As a result, while rough water-balance calculations can give an advance indication of management needs, the nature of the signal they give will always need further evaluation.

7 Conclusions: the way forward

7.1 Towards solutions: a process approach

The diversity and complexity of groundwater conditions, management needs, potential interventions and social contexts is a core theme running throughout this book. Conditions at local levels often differ in key aspects from trends observable at larger scales. Because of this diversity and scale issue, attempts to develop standardized solutions to groundwater problems that apply across large regions, or to specify the role of governmental and other institutions in advance, are likely to be ineffective. The basic principles of management listed in the preceding section apply but, in most if not all cases, they cannot be integrated into a single package of actions for uniform application. Instead it is important to develop a *process* that enables basic principles to be woven together in a manner that reflects local realities.

The process outlined in this section is intended to illustrate the form this might take in cases where governments or external agencies already recognize the need for management[18] and where there is no immediate threat to aquifer systems. The core idea relates to a phased approach. The approach starts with a commitment at policy levels to policy review and stakeholder participation (including establishment of the mechanisms necessary for that to occur) followed by establishment of a core group composed of stakeholders, experts, government representatives and concerned individuals. This group is essential in order to guide and move a long-term process forward. It then proceeds through analysis of needs and constraints to experimentation and, finally, to widespread implementation. Figure 11 attempts to illustrate this process.

[18] The authors make no claim that this process is applicable in all situations. In many cases, for example, the government may not recognize management needs, and external actors, such as non-governmental organizations, would need to drive the process and catalyse policy change. The goal of the section is to illustrate a potential approach which practitioners could use to spark their own thinking regarding the type of process that might work in their own particular setting.

Figure 11. A process to guide complex and diverse local implementation

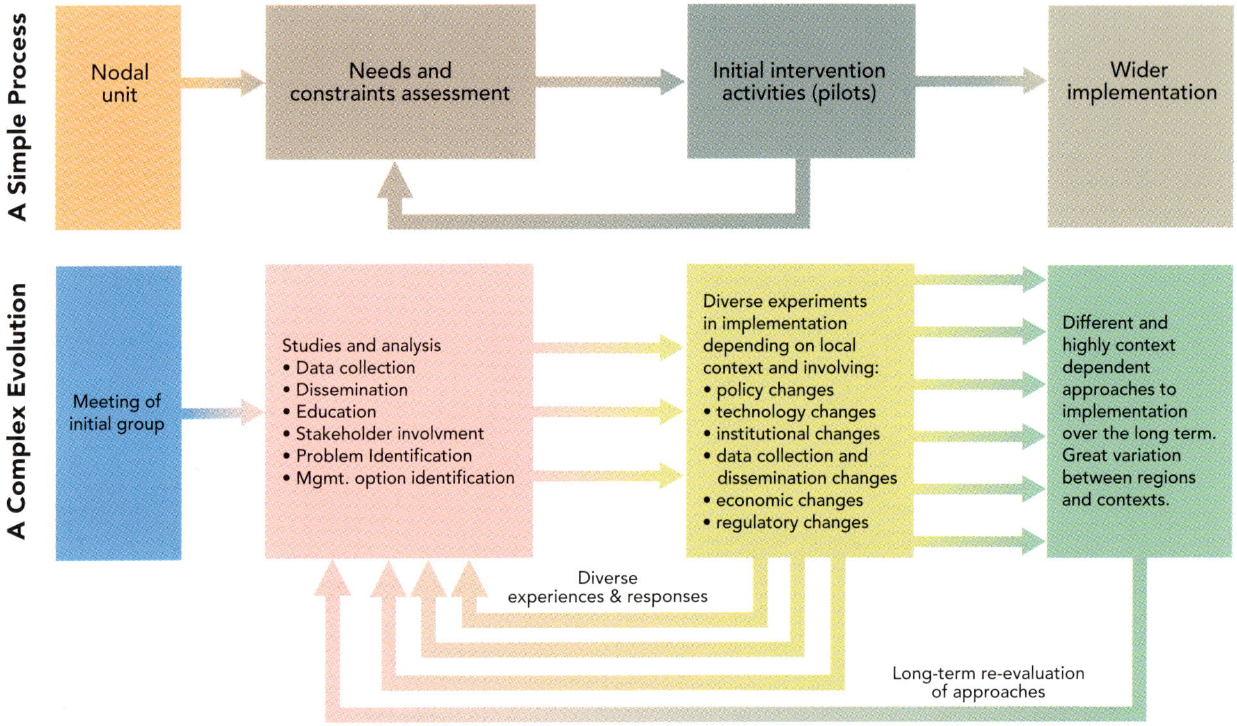

7.1.1 Phase I: identification or establishment of a nodal group

A first step towards the development of groundwater management systems or management institutions is for governments to make a firm commitment to policy innovation with stakeholder participation and to encourage formation of a group with sufficient interest and concern to guide and carry the process out. This will apply as much to surface water as to groundwater, but in the case of groundwater, the sensitivity to emerging groundwater problems needs to be well developed. This group could be based around an existing planning, policy or administrative unit. If so, however, it is important to ensure that it contains broad stakeholder representation. Communication, consultation and the establishment of formal roles for stakeholders are central to this. Consultation and communication can be started through such activities as seminars or workshops at the national and state levels that draw together individuals and organizations with interests or concerns regarding groundwater. Issues have to be clearly identified at this stage. The workshops can then be used as forums to initiate discussions of management needs and encourage the formation of a working group to guide the long-term process.

As the earlier discussion on stakeholder involvement indicates, membership is an important factor to consider in forming a working group or other nodal entity to guide the long-term process. In addition to a solid group of technical experts, working groups need to contain a broad spectrum of individuals, including water users, other stakeholders and academics. They also need to include individuals familiar with, and capable of, influencing legislation and the implementation of government policy. If these groups are not well represented, working groups tend to be dominated by individuals familiar with the technical aspects of groundwater management, and the process will tend to neglect the social and political considerations fundamental to success.

Nodal groups can sometimes be easy to identify, as is the case for remote communities in Dafur, Sudan.

7.1.2 Phase II: strategic analysis

Once a working group is formed, a strategic analysis exercise designed to produce a baseline assessment of needs, constraints and opportunities can be used as a tool to identify high-priority issues. This involves a broad review of:

- ***Groundwater services and social objectives.*** What are the key economic, environmental, social and other services provided by groundwater in the specific national and subnational contexts? What is the value of those services to society in economic and other terms?
- ***Groundwater problems.*** What are existing, emerging and foreseeable groundwater problems? In what way and to what extent do they threaten key services dependent on groundwater?
- ***Causes.*** What are the immediate causes and factors underlying the emergence of given problems? What factors appear to underlie the immediate causes?
- ***Data and understanding.*** Are data available to understand the resource base and its functioning sufficient to identify problems and management options? What are the key gaps in the scientific and technical information available?
- ***Stakeholders.*** Who are the important players who will need to be involved in any attempts to address specific problems or their root causes?
- ***Institutions.*** What are the roles and functions of current institutions and their suitability/capacity for addressing different aspects of emerging problems? Do existing institutions have the capacity to address management needs, including the involvement of critical stakeholders? If not, how might they be restructured or what capacities need to be developed for them to address those needs?

Completion of a needs and constraints assessment by the working group should be used to develop an initial strategic framework that specifies, at least on a conceptual level, the manner in which a broad array of interventions might be integrated in order to address groundwater problems. It should also result in the identification of a set of high-priority initial activities which can be implemented to move that overall strategic framework forward. In many cases, these might include:

- Initiation of a broad educational programme;
- Development of demonstration projects to gain experience with management issues and approaches;
- Research, data collection and analysis to address key scientific and policy information gaps;
- Initiation of dialogue with key stakeholder groups not included in the initial working group;
- Capacity-building exercises for existing institutions, with particular emphasis on strengthening systems for basic data collection and strengthening the social aspects of management.

Experience in similar review and strategy formulation exercises indicates that the above steps can be expected to take place over a minimum period of 2 to 3 years. The end result would be a much clearer understanding of management issues, needs and options in both the working group and society as a whole. In some instances, the needs may be much more urgent and rapid reviews have to be initiated to address urgent groundwater abstraction and pollution problems.

7.1.3 Phase III: starting to address fundamental challenges

By the third phase of the long-term process, sufficient experience should have been gained and a sufficiently broad base of stakeholders should be aware of management needs and options to begin to address more fundamental challenges. Dialogue over the complicated issues inherent in institutional and policy reform would be important to start at this phase. Representatives of stakeholders could be involved in such activities as the development of draft groundwater legislation and analysis of potential economic and regulatory instruments for management. Preliminary steps important for the development of groundwater management systems, such as the registration of wells, could also be appropriate at this phase. These new activities would complement expanded educational, capacity-building, research, data collection and pilot implementation activities started in the second phase. The focus in this phase would be to transform the initial conceptual strategic framework into a practical and implementable management and policy strategy.

Key products of this third phase would include the legislative and policy frameworks necessary for expansion of management initiatives beyond demonstration scales. To do this, the legislation would need to specify the process for initiating groundwater management activities in specific locations.

Whatever the process, legislation and policy changes would need to specify both the way management organizations can be formed at the scale needed and the array of powers they might utilize. A third phase of this type would probably require a number of years to implement. In most countries, major legislative and policy reforms occur only following a significant period of dialogue and debate. Furthermore, it is important for this debate and dialogue to occur. If, for example, legislation is passed but proves impossible to implement because of opposition from stakeholders who were not included in its development, the entire process can become deadlocked. Under these circumstances, the only option may be to initiate another long-term process of consultation in order to identify avenues for addressing opposition by key stakeholders. Avoidance of deadlock should, as a result, be a key objective. In most cases this will require identification of those stakeholders who have the political, social or economic capability to block effective action and ensuring that proposed approaches are acceptable to them.

7.1.4 Phase IV: wider implementation

A fourth phase in the long-term process would begin once some of the larger institutional and policy reforms are in place. This fourth phase would focus on a broad move from demonstration projects to widespread implementation. During this phase, a wide variety of policy changes and implementation activities would occur within a strategic framework that has been subject to extensive testing in earlier phases.

Taking progressive groundwater policy to scale is possible only if the institutional mechanisms are in place to enable natural resource management across communities, economic sectors and political or administrative jurisdictions. These are potentially complex linkages, but a degree of judgement has to be exercised in using the broad suite of management tools to induce change. Economic instruments, for example, may work well in sending groundwater conservation signals if communities have the latitude to adapt. They may fail—or lead to severe social disruption—if reductions in water consumption cannot be achieved or there is no system of accountability.

7.2

Principles in practice: case studies

How the above basic principles and process might translate into action will vary greatly among locations. The authors have selected a set of case studies to illustrate the variety of responses and check the applicability of the principles (annex). Mexico and Barbados are examples of countries starting on this long-term process of developing the basic information, capacity and systems necessary to manage their groundwater resources. In contrast, the American High Plains is an example of a more evolutionary process, through which groundwater management and the institutions required to implement it have gradually evolved in the absence of any integrated national or regional policy initiative. These examples of groundwater management are in a minority. Groundwater management per se is more typically left as an afterthought in water management policy reviews and has little influence in determining the character of rights systems or the spatial aspects of water resource planning. In many cases it is now too late to address groundwater problems effectively. The current inertia in Pakistan, where the sustainability of the Indus basin hinges on addressing groundwater-related problems, is a case in point (World Bank, 1994). The inability to comprehensively manage and protect the prolific karstic aquifers in Zambia is also a prime example of policy failure and has resulted in the reliance on expensive surface-water schemes (Burke, 1994) and inadequate protection of productive but vulnerable aquifers. In some countries, such as India, there are the beginnings of movement. Recently, China has attempted to bring groundwater into the realm of water resource management through major institutional reform at the centre. This is, however, far too late for many of the principal aquifers in north-eastern China, which already exhibit

The principle of groundwater as a source for potable supplies for rural centres in the Democratic Republic of the Congo was implemented even when raw surface water was plentiful.

substantial water-level declines (British Geological Survey, 1994). Nonetheless, the examples cited below provide a measure of what is possible in three widely differing circumstances. The High Plains case is given particular attention because the historical evolution of groundwater institutions has occurred over a much longer time period than in most other regions. In addition, the case illustrates many of the practical interventions central to groundwater conservation in agriculture and the core institutional and social factors underlying their adoption. Finally, the High Plains case demonstrates that *effective action can occur in situations where groundwater rights have not been quantified or allocated to individuals and where management agencies lack regulatory powers.* This final point is of great importance because attempts to manage groundwater in many parts of the world are deadlocked over politically sensitive legal and rights reform issues. While legal and institutional reform is often needed, the High Plains case indicates that approaches based on economics, education and technical innovation (spheres of intervention that are generally far less politically sensitive than legal reform) can address management needs in some situations.

7.2.1 Some lessons drawn from the case studies

The case studies illustrate key issues related to groundwater that are common throughout the world. Groundwater management needs cannot be viewed as "problems" that, once "solved", require little attention. Instead, current problems signal the emergence of management needs that will require long-term attention. Barbados has had strictly enforced groundwater regulations in place for over 30 years, but increasing demands on the resource base from growing populations and the resource's increased vulnerability to pollution necessitate the introduction of new—and often expensive—interventions if water supplies are to be protected. Management in Barbados, as in Mexico, is an ongoing process. Mexico has outlined an ambitious programme for the development of management capacity, covering the period extending to the year 2010. The National Water Law laid the conceptual foundations for this process and the Government has begun the hardest work of identifying and regulating groundwater extraction concessions. Data and other support systems are also being built. The process will, however, not end in 2010. If all goes well, those working on water management in Mexico hope that users will have agreed to specific management actions by the year 2007. They also recognize that achieving this will require very intensive education and capacity-building. As a result, even under the best of circumstances, by the year 2010 communities will just be starting to implement the actual process of groundwater management. The inevitable array of new and difficult issues associated with implementation will only just be beginning to emerge.

The High Plains case and the lessons it contains are quite different from the Mexico and Barbados cases. It illustrates that locally governed management institutions can take effective action even in the absence of regulatory powers or unified policy and administrative frameworks. Groundwater districts in the High Plains are legally recognized and require this recognition in order to take action or develop programmes. They do not, however, follow any single set of regulatory, administrative or other management approaches. In some cases, such as Texas, management entities have little regulatory authority. There, the only limitation on individual groundwater rights is a provision against "waste" and, consequently, management entities cannot rely on regulatory powers to achieve management objectives. Furthermore, management does not address an entire hydrologic unit. The Ogalalla is a regional aquifer system, but management, which in many cases effectively addresses key groundwater problems, occurs at a much more local level. Beyond this, the High Plains case provides key insights into the human factors underlying successful management. Farmers in the High Plains are adopting efficient irrigation practices primarily for economic reasons rather than just to save water. Furthermore, many of the practices that are most successful involve changes in the care with which water is actually used rather than changes in technology. The importance of irrigation scheduling illustrates the key role micro-management by users plays in determining the overall efficiency of water use, regardless of technology. This leads to the question of participation. By many measures, groundwater management in the High Plains is not "participatory" in that few users are actively involved in the day-to-day operations of groundwater institutions. At the same time, the institutional structures do ensure that policies reflect popular preferences within the management area. This indirect participation appears to be central to management in the region. Finally, the High Plains case illustrates a key issue in the goals management may address. Because recharge is low, in most cases management in the High Plains focuses on planned depletion rather than sustainable maintenance of the groundwater resource base. Sustainability would, in many areas, require reducing use to the point where groundwater could no longer serve as a major resource for irrigated agriculture. A planned depletion approach seeks to enable orderly transition of the regional economy as groundwater resources gradually decline.

For some communities, such as those subsisting in the Tihama Plain, Yemen, options for groundwater management will remain very limited.

7.3 Conclusions

The core message of this review of groundwater and society is that emerging and existing groundwater problems threaten a wide array of services—from food security and clean drinking water to instream flows and other environmental features. These services are of fundamental importance to human society. For this reason, social dimensions are as important as technical dimensions in the development of approaches to address these problems and ensure the sustainability of key social, economic and environmental systems. Basic data and information on aquifer status and the projected demands, as well as effective user participation based on education and constructive dialogue, are essential inputs for resolving complex issues related to groundwater.

In many respects, groundwater poses a distinct set of technical and socio-economic challenges. It is not safe to assume that aquifer systems can be managed with the same set of tools, analysis and institutional solutions that are deployed for the management of surface water. In many cases, the intimate and intense engagement with the resource base by millions of individual users precludes the use of the more centralized command and control management approaches that have been developed for river basin management. Furthermore, there is little evidence that the inherent variability of the resource and the socio-economic response to the opportunity it presents have been fully incorporated into contemporary water management approaches. *Management needs and options vary greatly among regions.* While it may be possible to generalize the process of identifying emerging problems and management options, actual management approaches and interventions need to be tailored to local contexts.

The need for a strategic framework to address this level of local complexity has been articulated. While this is a conceptual tool, it is designed to prompt an approach to groundwater management that accurately reflects the spatial, temporal and sociocultural dimensions of groundwater use and avoids purely technical or administrative solutions. The evidence from many of the country examples cited in this review is that such narrow approaches have failed. The issue of time is fundamental. Damage to aquifers has occurred over a relatively short time period with the spread of energized pumping technologies and "modern" pollutants. Management responses are often late and in many cases are limited to damage control rather than preventive care. It is also important to recognize that, whatever the urgency of management, it often cannot be implemented in practice without broad social support and understanding. This suggests a need to be strategic in identifying and implementing solutions, but equally not to be impatient in expecting positive results. Over the long term, education of children is particularly important in order to shift society's awareness and understanding of both groundwater problems and management opportunities. Groundwater problems are generally not amenable to rapid solutions. Rather, they signal the need to develop management systems capable of addressing and predicting constraints in a flexible fashion, as they arise. In many developing countries this process is in its infancy, and it is crucial to initiate and encourage whatever management actions are currently feasible while starting the long-term process of developing flexible, integrated management systems. *The evolution of groundwater management systems is a long-term process.* If it is unscientific to expect aquifer systems to respond immediately to technical management, it is equally unrealistic to expect the dependent social systems to respond over short periods of time.

It is always important to remember that certain communities will never have rights in land and water.

Refugee camp, Baluchistan, Pakistan

Annex: case studies

Preface

The case studies are drawn from examples submitted by members of the expert group. They illustrate the range of continental and small island groundwater issues that have been addressed and how solutions can be adopted.

Water rights administration for groundwater management in Mexico[19]

The Mexican Constitution has, since 1917, declared that water is common property and that surface water can be used only if a concession is issued by the executive branch of government. Groundwater can be freely abstracted except in areas where the executive branch declares government control, under which concessions are required. At present there are approximately 300,000 concession holders in Mexico. These include Mexico City, with 17 million people, several hundred irrigation modules of 1,000 to 5,000 hectares (each with tens or hundreds of individual users), industries, medium-sized cities and small villages with a few houses.

The National Waters Law, approved by Congress in December 1992, calls for integrated management of the quantity and quality of surface water and groundwater within watershed boundaries and establishes the National Water Commission as the sole federal water authority. In addition to existing concession requirements, the law requires permits in order to discharge waste water into national waters and establishes other obligations and rights of water users. These rights and obligations make up the core *regulatory tools* established under the law to manage water resources. The law also establishes *economic tools*, including levies for abstraction and for waste-water disposal, and the possibility of trading water-use rights (the concessions) under given conditions. Finally, the law establishes *participation tools* in the form of watershed councils. The councils are intended to contribute to water resource planning and management in watersheds and aquifers, where over-abstraction, pollution or flooding problems demand the coordinated action of federal, state and local authorities, along with users and stakeholders. A delicate bal-

[19] By Hector Garduno, National Water Commission, Mexico.

ance of these three tools is required, but no reasonable water rights trading can be enforced before water availability is thoroughly assessed, and water uses are reliably known and the corresponding rights are adjudicated.

Mexico has an annually renewable water resource base of some 470 km^2, of which approximately 40 km^2 is furnished by groundwater. The estimated total abstraction of surface water and groundwater for consumer use is 87 km^2 per year, 76 per cent of which is for agriculture and livestock. Of the yearly total, 28 per cent is groundwater. This is drawn from 459 identified aquifers, 80 of which are estimated to have annual abstraction rates that exceed replenishment by more than 20 per cent. In some locations, over-abstraction is causing severe land subsidence, saline intrusion and migration of pollutants.

Passage of the National Waters Law set the stage for sustainable water resource management. Sustainability cannot, however, be achieved overnight. It has been estimated that the initial implementation programme shown in the table below will require 15 years and will involve a complex set of activities.

Despite the long-term nature of the process, progress to date has been substantial. Between 1917 and 1992, the power to issue concessions was centralized under only a few civil servants, and only 2,000 concessions were issued because priority was given to water resource development, not to management. Enforcement of the new water law was started in 1993 by empowering 40 officials all over the country to issue concessions. In addition, Presidential decrees were issued outlining conditions and benefits for irregular users who apply for their concessions before December 1998. By November 1997, almost 183,000 users had applied and 66,000 had been regularized. The Presidential decrees are based on the honesty of users. Concessions are issued for 10 years[20] to those users who, under oath, declare they were abstracting water prior to October 1995, when the decrees were published. The National Water Commission can check this in the field and will apply severe penalties to those who have not been truthful about either the volume or the date of the initial abstraction.

To date, groundwater technical committees (GWTCs) have been established in two aquifers to facilitate user participation with federal, state and local authorities in designing and implementing groundwater management regulations. It is expected that by the end of 1998 most users will have obtained abstraction concessions and waste-water permits. It is also expected that by mid-1999 watershed councils will have been established in problematic watersheds and that by 2000 GWTCs will have been established in sensitive aquifers.

At the same time, a Modernization of Water Resources Management Programme, with partial financing from the World Bank, is expected to improve the quantity and quality of databases for surface water and groundwater by 2001. It is therefore realistic to expect that by 2007 users will have agreed and put in place regulations to restore the hydrologic balance of groundwater, cope with surface-water variability and set up realistic quality goals for aquifers, rivers and lakes. This will, however, be achieved only if intensive programmes to raise awareness and build capacity are simultaneously implemented for all players in the water sector.

---

20 According to the law, concessions may be issued for periods from 5 to 50 years. Given that water uses are not all known and that consequently it is difficult to compute the net available water, a 10-year period was selected to allow for a transition, in order to register most users in the Water Rights Public Registers.

Protection of peri-urban supplies: Bridgetown, Barbados[21]

The Caribbean island of Barbados is almost totally dependent on groundwater for public water supply. The population of 260,000 relies on water drawn from 17 production wells, which tap a jointed, porous and karstic unconfined aquifer in limestone formations similar to those that occur widely in the Caribbean, Florida and the Gulf of Mexico region. As an island with this type of aquifer, Barbados's groundwater supplies are highly vulnerable to pollution and saline intrusion. Flow rates throughout much of the aquifer are rapid, and pollutants or salinity could rapidly contaminate most available supplies. A safe and reliable water supply from the island's aquifer is vital for economic as well as public health reasons. In addition, hotels serving tourism, the mainstay of the Barbadian economy, draw their water from the public supply system.

Figure A1 Map of development control zones in Barbados with summary of principal features

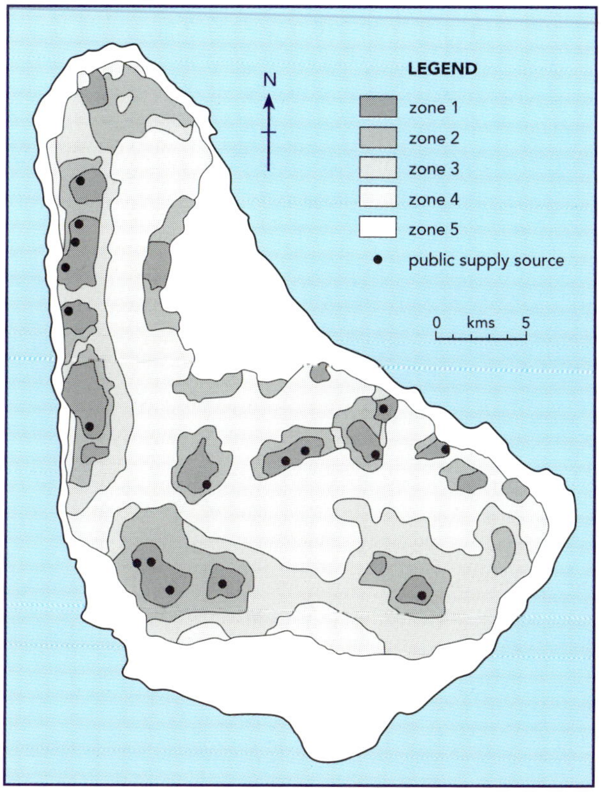

Principal features of Development Control Zones in Barbados

Zone	Definition of outer boundary	Maximum depth of soakaway pits*	Domestic controls	Industrial controls
1	300-day travel time	None allowed	No new housing. No changes to existing waste-water disposal.	No new industrial development.
2	600-day travel time	6.5 m	Septic tank of approved design, discharge to soakaway pits. Separate soakaway pits for toilet effluent and other domestic waste-water. No storm run-off to sewage soakaway pits. No new fuel tanks.	
3	5- to 6-year travel time	13 m	As above for domestic waste-water. Fuel tanks to approved leak-proof design.	All liquid industrial waste to be dealt with as specified by Water Authority and maximum soakaway pit depths as for domestic waste.
4	Extends to all high land	No limit	No restricitions on domestic waste-water disposal. Fuel tanks to approved leak-proof design.	
5	Coastline	No limit	No restrictions on domestic waste-water disposal. Siting of new fuel storage tanks subject to approval of Water Authority.	

*Soakaway pits are excavations down to fissures, designed to provide most rapid disposal of water.

[21] By Brian Morris, British Geological Survey, Wallingford, United Kingdom.

The extremely high vulnerability of this aquifer has been recognized by the Government of Barbados for many years, and in 1963 it passed important and far-sighted legislation implementing a policy of development control zones around existing and proposed public supply sources. These were based on simplified concepts of pollutant travel times through the aquifer. Domestic or industrial activities at the land surface within each zone are subject to a hierarchical system of controls that are increasingly strict as travel time decreases (see figure A1). The island's groundwater protection policy also includes saline intrusion prevention measures, through careful design and control of abstraction regimes at each well, and both elements of the policy have been firmly enforced for more than 30 years despite increasing development and population pressures. A detailed study during the period 1987-1990 (Chilton et al., 1990) assessed pollution risk from intensification of agriculture, uncontrolled discharge of industrial effluents and on-site sanitation. It concluded that the system of development control zones had served Barbados well in protecting the groundwater resources on which it depended, despite a number of potential risks needing additional controls. These risks arose because of changes in agriculture from sugar cane to cash crops, encroachment of the Bridgetown suburbs further into groundwater catchments and a rise in small-scale industries. The need to protect the groundwater resource is a high priority for the island, reflecting the high cost and technical difficulty of alternative supplies.

The core water-supply and disposal problems in Barbados are thus:

- The highly vulnerable phreatic aquifer, in which groundwater flow is rapid;
- The total dependence on groundwater for public supply and the important tourist industry;
- The shift in the island economy from the sugar-cane monoculture to the intensification of horticulture/agriculture;
- The ubiquity of on-site sanitation;
- The high cost of substituting other sources for supplies lost from contamination of the aquifer;
- The very high value of groundwater.

In this context, some of the water management options open to Barbados include:

- Development of additional groundwater sources in already protected, still rural, areas;
- Improvement in the distribution system and leakage reduction;
- Extension of sewered waste-water systems to the most sensitive peri-urban control zones around Bridgetown;
- Maintenance of high treatment standards to safeguard public health and beaches for the tourist industry.

Western United States, High Plains[22]

Introduction

The story of the American High Plains is very much the story of water. Changing perceptions of the region express prevailing views of water availability. Known as the Great American Desert and later the Dust Bowl, the region gained settlers whenever abundant rain would fall for a few years. The variability of the region's weather is revealed by these differing views and corresponding periods of in- and out-migration.

Stability in farming and occupancy was sought in irrigation. Early irrigators diverted water from rivers, but limited surface flows and precipitation restricted irrigated land to a few thousand hectares. An irrigation boom in Texas based on groundwater began during the 1930s. After the Second World War, irrigation spread rapidly through portions of Colorado, Kansas, Nebraska, New Mexico, and Oklahoma underlain by groundwater. It became known as the Land of the Underground Rain (Green, 1973).

The regional economy and population boomed, but the good times might not be sustainable. The water was mostly being pumped from the High Plains aquifer, which receives minimal recharge away from sandy areas. Many wells did go dry, some farmers no longer could irrigate their crops, and a few towns had to haul in water as heavily used parts of the aquifer were depleted.

The High Plains aquifer

The lifeblood of the High Plains economy is the High Plains regional aquifer. It sustains 20 per cent of the irrigated area within the United States. The High Plains regional aquifer comprises several water-bearing formations. Most important is the Ogallala, underlying about 347,000 square kilometres. Other geologic units in the High Plains aquifer include the Brule formation, the Arikaree group and the deeper Dakota formation (Sophoceous, 1998). Saturated thickness, specific yield and areal extent of the aquifer are not uniform. Nebraska has almost two thirds of the High Plains groundwater, followed by Texas with 12 per cent, Kansas with 10 per cent, Colorado with 4 per cent and Oklahoma with 3.5 per cent. New Mexico, South Dakota and Wyoming have less than 2 per cent each (Kromm and White, 1990).

Because of low precipitation and high evapotranspiration, groundwater recharge is negligible throughout most of the High Plains. Estimated recharge rates range from a high of 15 centimetres per year in sandy areas of Nebraska and south central Kansas to just 0.06 centimetre per year in parts of Texas. The long-term average recharge rate is probably less than a centimetre per year for the High Plains region (Gutentag et al., 1984). The United States Geological Survey estimates that the area-weighted average decline in the High Plains aquifer was about one-half metre from 1980 to 1994, compared to over 3.2 metres from pre-development (1940) to 1980. Since 1980, the annual area-weighted water-level decline is about 0.36 mm, as compared with 82 mm between 1940 and 1980 (United States Geological Survey, 1995).

---

[22] David E. Kromm, Department of Geography, Kansas State University, Manhattan, KS 66506-0801. krommgeo@ksu.edu

The economics of groundwater exploitation are determined in part by energy costs, which increase with greater depth to the water-table. An irrigator who pumps water from 30 metres below the ground uses only about one third the energy of an irrigator who pumps from a depth of 90 metres. Survey research (Taylor et al, 1988; Kromm and White, 1990) suggests that irrigators throughout the High Plains are more concerned about high energy costs than about groundwater depletion. In addition, well yields decrease in relation to declining water levels according to an inverse-square relationship. For example, a 50 per cent reduction in saturated thickness could mean that a well will yield only 35 per cent of its initial capacity. Generally, if the saturated thickness is less than 10 metres, the remaining water is not economically recoverable (Sweeten and Jordan, 1987).

Irrigated agriculture

Irrigated agriculture sustains the High Plains. It forms the keystone of an integrated agribusiness economy that demands seeds, fertilizers, pesticides, agricultural machinery and credit. Without irrigation, vast tracts of land now cultivated would be in pasture or extensively farmed with dryland techniques and the regional economy would be much smaller and far less active (Kromm and White, 1992a).

Irrigated agriculture is not a single static activity in the High Plains. Crop choice, irrigated methods, groundwater management and amount of land irrigated vary. Statistics on irrigated area peaked in the 1978 census year, when just over 5,580,000 hectares were cultivated (table A1). The actual greatest irrigation probably occurred two or three years later. Declines in irrigated area have been accompanied by a shift northward. New Mexico and Texas contributed their highest proportions in 1959 and Oklahoma in 1978. Colorado, Kansas and Nebraska had their greatest proportions of regional irrigated hectares in 1987. Nebraska became increasingly important, expanding its proportion of irrigated agriculture from 27.1 per cent in 1969 to 37.9 per cent in 1978 and rising to 46.8 per cent in 1987. All six states saw a decline in irrigated land after 1978, with Texas experiencing the greatest loss and Nebraska the least (United States Census of Agriculture, varied years).

Declines in irrigated area have not been steady. Nearly 600,000 irrigated hectares were added between 1987 and 1992, an increase of almost 12 per cent, and expansion may have continued throughout the 1990s. There have also been substantial changes in irrigated hectares for specific crops (table A2). In recent years, the proportion of total irrigated hectares accounted for by corn and wheat has increased significantly. These changes in crops reflect expansion of the feedlot and beef-processing industries in the High Plains, along with federal crop subsidy programmes that have favoured corn.

Table A1 Total irrigated hectares, United States High Plains, 1949-1992

STATE	1949	1959	1969	1978	1987	1992
Nebraska	330,409	804,695	1,095,338	2,118,181	2,106,148	2,351,512
Texas	670,348	1,652,368	1,951,485	1,973,634	1,156,192	1,394,535
Kansas	51,407	229,903	515,731	840,853	721,750	798,418
Colorado	83,119	111,548	223,433	387,608	318,204	335,002
Oklahoma	2,672	23,781	116,435	129,946	104,141	114,575
New Mexico	54,117	101,600	128,244	131,606	97,239	103,645
Region	1,192,072	2,923,895	4,030,666	5,581,828	4,503,674	5,097,687

Source: U.S. Census of Agriculture, 1949, 1959, 1969, 1978, 1987, and 1992

Table A2 Irrigated hectares by crop, United States High Plains, 1969-1992

CROP	1969	1978	1987	1992
Corn	1,119,438	2,506,819	2,111,876	2,597,081
Wheat	444,268	526,401	582,515	687,344
Sorghum	1,047,686	605,012	458,737	514,585
Hay	208,501	342,766	344,013	335,914
Cotton	563,149	671,033	434,324	238,332
Other *	647,624	929,797	572,209	724,431
TOTAL	4,030,666	5,581,828	4,503,674	5,097,687

*Includes soybeans, oats, pasture, and some of listed crops not disclosed.
Source: U.S. Census of Agriculture, 1969, 1978, 1987, and 192

On-farm water management

Management of the Ogallala aquifer begins with the farmer, as more than 90 per cent of the water pumped is used for irrigation. In the early years of irrigation, many farmers thought of the High Plains aquifer system as an endless supply of water connected to Rocky Mountain snowmelt. Even as the finiteness of groundwater became accepted, there was "a kind of secular faith that technology is infinitely capable of solving future problems involving overdraft of the Ogallala" (Green, 1992, p. 42). Resource and technology limitations are much better understood today, and varied management strategies are employed to conserve groundwater. Irrigators recognize that they are "the key agents in any program to extend the life of the aquifer, and most irrigation farmers take the responsibility seriously" (Kromm and White, 1992b, p. 226).

Irrigators are adopting a wide assortment of water-saving practices. Many of the technologies to conserve water in the High Plains are generally available. Kromm and White have found that location explains regional variability in adoption better than socio-economic characteristics, farm variables or hydrologic characteristics (Kromm and White, 1990). As irrigators do different things at different locations, it is clear that more work is needed to understand the diffusion of water-saving practices, so that the information can be distributed uniformly, the optimal mix of practices adopted and conservation efforts encouraged (Kromm and White, 1991).

What prompts efforts to conserve groundwater? Depth to water is more strongly associated with whether irrigators adopt water-saving practices than the amount of water as measured by saturated thickness of the aquifer (Kromm and White, 1990). Conservation efforts result more from the high costs of energy than from fear of the decline of an aquifer. Irrigators who pump water from over 80 metres and pay higher energy bills are more likely to adopt water-saving practices than those who have less water that is nearer the surface. Past research has found that a majority of irrigators "adopted water-saving practices to conserve energy, to reduce labor, and to increase yields or when they had to replace existing equipment—not just specifically to save water although saving water was viewed as a positive advantage" (White and Kromm, 1995, p. 227).

Water-saving practices can be divided into three categories: field practices, management strategies and system modifications. Field practices are such techniques as chiselling compacted soils, furrow dyking to prevent run-off, and levelling land, which keep water in the field, distribute it more efficiently or encourage soil moisture retention. They generally do not require high capital investment. Management strategies monitor soil and water conditions and collect information to assist scheduling and irrigation system improvement decisions. System modifications, such as adding drop tubes to centre pivots, retrofitting wells with smaller pumps, installing surge irrigation or constructing a tailwater recovery system, are usually expensive (Kromm and White, 1990). There is a wide range of system efficiencies depending upon type, design and management. Table A3 presents a comparison of performance ranges by system type. In the High Plains, the leading form of application is the centre-pivot sprinkler system, followed by furrow irrigation, often using surge valves. Recently, water management has improved marked-

Table A3 Range of application efficiencies for various irrigation systems

System type	Application efficiency range (per cent)
Surface irrigation	
Basin	60-95
Border	60-90
Furrow	50-90
Surge	60-90
Sprinkler irrigation	
Handmove	65-80
Traveling gun	60-70
Center pivot and linear	70-95
Solid set	70-85
Microirrigation	
Point source emitters	75-95
Line source emitter	70-95

Source: Rogers et al., 1997

ly through irrigation scheduling. Irrigation scheduling involves determining when to irrigate and how much water to apply in terms of the moisture demands of the crop. This management intervention has been one of the most important factors in water conservation on the High Plains. Micro-irrigation in the form of subsurface and drip irrigation is taking hold in some areas and achieves even higher water application efficiency. Because of costs and other considerations, adoption of water conservation techniques is variable. In addition, there is not a common knowledge base with respect to water-saving methods on the High Plains. Groundwater management districts cover only some of the area, and the programmes of these organizations vary significantly.

Water management institutions

Institutions play a major role in High Plains irrigation. Water rights are administered by various levels of organizations. Water falls principally under state control, but substate or local districts are common. These local authorities are formed usually to ensure that existing rights, often for irrigation, are protected. The authority of the local districts varies from largely educational to requiring land-use planning and metering of irrigation water use. Most of the states now have water or resource management districts that are specifically responsible for the Ogallala. Although planned and orderly depletion is the most common goal, one groundwater management district in western Kansas is considering a zero-depletion policy.

In the High Plains Ogallala region, four states (Colorado, Kansas, Nebraska and Texas) have attempted to take a local decision-making approach to groundwater management in order to cope with groundwater depletion. Groundwater management districts in Colorado and Kansas have greater authority than the underground water conservation districts in Texas or the natural resource districts in Nebraska. Direct state control prevails in New Mexico and Oklahoma. Local control began in Texas almost 50 years ago. The largest district, which is headquartered in Lubbock, has been the most active in the High Plains in advocating and assisting in on-farm irrigation efficiency. In 1969, Nebraska combined resource agencies dealing with the management of soil, water and wildlife into 26 regional natural resource districts (Templer, 1992).

In both Colorado and Kansas, districts have broad management authority to include recommending the rejection of requests for new wells, requiring well metering, well spacing and pumping limitations, development of management plans, assessing special taxes and issuing bonds to finance irrigation systems. In both states, districts are governed by a board of directors elected by the public at either a general election (Colorado) or a widely publicized groundwater management district annual meeting (Kansas). Any adult landowner can serve on the boards in Colorado, while Kansas also requires a minimum holding of 40 contiguous acres. However, variation in the devolution of water management responsibility is significant. In Kansas, local districts have demanded great autonomy, whereas those in Colorado primarily carry out state policies. In Kansas, each district includes several counties, whereas in Colorado most fall within a single county. As a rule, these districts correspond with the occurrence of the Ogallala aquifer and an intense level of irrigation (White and Kromm, 1995).

The general objectives of the groundwater management districts in Colorado and Kansas are stipulated in the enabling legislation of both states, and individual districts have adopted clearly stated rules, regulations and programmes. All districts regulate well spacing, new well development, abandoned wells, and points of water diversion. Regulations governing new wells are generally obsolete because the districts are closed to further irrigation development. Regulations governing abandoned wells require that they be handled to prevent groundwater pollution and hazards to individuals and animals. When ambiguity exists as to what local districts are authorized to do, state courts have usually ruled in favour of the local districts. In Colorado, none of the districts exercise the full range of authority made possible by the enabling act (White and Kromm, 1995).

Public participation

Groundwater management districts in the High Plains aquifer region were formed to bring about local control of groundwater management. The districts' enabling acts empowered local people to take a more active role in their destiny. By having locally elected boards establish policy, the districts were to avoid the bureaucratic indifference and professional biases often seen in state and federal agencies. The operations of the districts differ markedly, suggesting responsiveness to local preferences and conditions. Although supporters of the groundwater management districts believe that they provide true democratic participation, there remains the question regarding whose interests are being favoured and whose are being ignored (White and Kromm, 1995).

If the districts are to be effective grass-roots organizations, their initiatives and policies should come from people and reflect popular preference so far as groundwater use is concerned. Research projects throughout the entire High Plains have found that local preferences generally agree with local district priorities. Most respondents favoured their districts' programmes over policies pursued elsewhere. Overall, groundwater districts appear largely to reflect local popular will. That said, there is little direct public participation in groundwater management district operations. Few people appear at regularly scheduled board meetings or more widely publicized meetings held for public input. Not many cast votes in elections for board members. Conflicts seldom occur, with consensus being the mode of operation. Public participation is weak, but popular will appears to be served (White and Kromm, 1995).

Conclusion

Capital-intensive, very productive High Plains agriculture can be thought of as an experiment in irrigation, as the water source is finite and being depleted. Although significant improvement in irrigation efficiency has been achieved and localgroundwater management institutions have become increasingly effective in encouraging conservation, it remains uncertain if the largest irrigation development in the western hemisphere will survive more than two or three more generations.

Bibliography

Acworth, I. A. (1987). The development of crystalline basement aquifers in a tropical environment. *Quarterly Journal of Engineering Geology*, 20, 265-272. Geological Society, London.

Agarwal, A., and S. Narain, eds. (1997). *Dying Wisdom: Rise, Fall and Potential of India's Traditional Water Harvesting Systems.* New Delhi, Centre for Science and Environment.

Alderwish, A. M., and J. Dottridge (1998). "Recharge components in a semi-arid area: the Sana'a Basin, Yemen." In Robins N. S. (ed.), *Groundwater Pollution, Aquifer Recharge and Vulnerability.* Geological Society, London, Special Publications 130, 169-178.

Al-Eryani, M. I. (1995). Legal Feasibilities of the Various Options for Sana'a Water Supply. Sana'a, Extended Coordinative Meeting on the Future for Sana'a Water Supply.

Alley, W. M., ed. (1993). *Regional Groundwater Quality.* Van Nostrand Reinhold, New York. 634 pp.

American Society of Civil Engineers (1961). *Groundwater Basin Management.* American Society of Civil Engineers Manual of Engineering Practice, No. 40.

Appleton, J. D., et al., eds. (1996). Environmental geochemistry and health with special reference to developing countries. Geological Society Special Publication No. 113. The Geological Society, London.

Bachu, S., and J. R. Underschultz (1993). Hydrogeology of formation waters, Northeastern Alberta Basin. *AAPG Bulletin* (77)10, 1745-1768.

Baker, M., and J. Romm (1990). *Emerging Institutional Linkages between Land and Water: Opportunities for State Involvement in Watershed Management Institutions.* Berkeley, University of California.

Barraqué, B. (1996). Eurowater. Water Rights and Administration in Europe. EU, D.G.XII. Horizontal Report No. 8.

Barraqué, B. (1997a). *Les Agences de l'Eau et la Question du Patrimoine Commun en France et en Europe.* Commissariat Général du Plan. Service Energie Environment/Agriculture. Secteur Tertiare.

Barraqué, B. (1997b). *Groundwater Management in Europe: regulatory, organizational and institutional change.* How to cope with degrading groundwater quality in Europe, FRN-NFR MAB-UNESCO-IHP. Stockholm.

British Geological Survey (1994). Aquifer Overexploitation in the Hangu Region of Tianjin, PRC. British Geological Survey Technical Report WC/94/42R. Keyworth, Nottingham, United Kingdom.

Bristish Geological Survey (1995). Overexploitation of Aquifers—Final Report. British Geological Survey Technical Report WC/95/3. Keyworth, Nottingham, United Kingdom.

British Geological Survey (1996). Unconsolidated Sedimentary Aquifers Review No. 14. Groundwater Modelling in Aquifer Management. British Geological Survey Technical Report WC/96/64. Keyworth, Nottingham, United Kingdom.

Blaikie, P., and T. Cannon (1994). *At Risk.* London, Routledge.

BOSTID (1986). *Proceedings of the Conference on Common-Property Resource Management.* Common-Property Resource Management, Washington, D.C., National Academy Press.

Brown, L. R. (1970). *Seeds of Change: The Green Revolution and Development in the 1970s.* New York, Praeger.

Burke, J. J. (1994). Approaches to integrated water resource development and management: the Kafue Basin, Zambia. *Natural Resources Forum* 18(3): 181-192.

Burke, J. J. (1996). Hydrogeological provinces in central Sudan: morphostructural and hydrogeomorphological controls. *Groundwater and Geomorphology.* T. Brown. Chichester, Wiley.

Caponera, Dante A. (1996). *Principles of Water Law and Administration.* National and International. Balkema, Rotterdam, 260 pp.

Central Ground Water Board (1996). Background Note Prepared by Central Ground Water Board. *Colloquium on Strategy for Ground Water Development,* New Delhi, Government of India, Ministry of Water Resources.

Chaturvedi, M. C. (1976). *Second India Studies: Water.* New Delhi, Macmillan Company of India Limited.

Chilton P. J., et al. (1990). A ground-water pollution risk assessment for public water-supply sources in Barbados. In: Krishna J. H. l (ed.). *Proceedings of the International Symposium on Tropical Hydrology and Fourth Caribbean Islands Water Resources Congress,* Puerto Rico, July 22-27, 1990.

Chilton. P. J., ed. (1997). Groundwater in the Urban Environment. Volume 1. *Problems, Processes and Management.* A. A. Balkema, Rotterdam.

Ciriacy-Wantrup, S. V., and R. C. Bishop (1975). "Common property" as a concept in natural resources policy. *Natural Resources Journal*, 15: 713-727.

Dains, S. R., and J. R. Pawar (1987). Economic Returns to Irrigation in India. New Delhi. Report prepared by SDR Research Groups Inc. for the United States Agency for International Development.

Department of Water Resources (1994a). The California Water Plan Update, Executive Summary. Sacramento, Department of Water Resources, State of California.

Department of Water Resources (1994b). California Water Plan Update. Volume 1. Sacramento, Department of Water Resources, State of California.

Dhawan, B. D. (1995). *Groundwater Depletion, Land Degradation and Irrigated Agriculture in India.* New Delhi, Commonwealth Publishers.

Dubourg, W. R. (1997). Reflections on the meaning of sustainable development in the water sector. *Natural Resources Forum*, 21(3): 191-200.

Edmunds, W. M., and P. L. Smedley (1996). Groundwater Chemistry and Health: an overview. In Appleton, J. D., Fuge, R. and McCall, G. J. H., eds. (1996), Environmental geochemistry and health with special reference to developing countries. Geological Society Special Publication No. 113. The Geological Society, London.

Edmunds W. M., and Wright, E. P. (1979). *Groundwater recharge and paleo-climate in the Sirte and Kufra Basins.* Libya. J. Hydrol (40).

Edmunds, W. M., et al. (1987). Estimation of aquifer recharge using geochemical techniques. Final Report of the Lower Atbara River Basin Project. BGS/NERC Report WD/OS/87/1.

Esrey, S. A., and J. P. Habicht (1986). Epidemiologic Evidence for Health Benefits from Improved Water and Sanitation in Developing Countries. *Epidemiologic Review* 8: 117-128.

Food and Agriculture Organization (1997). Irrigation in the Near East Region in Figures. Water Report 9. Food and Agriculture Organization of the United Nations. Rome.

Gleick, P., ed. (1993). *Water in Crisis: A Guide to the World's Fresh Water Resources.* New York, Oxford University Press.

Gleick, P., et. al. (1995). *California Water 2020: A Sustainable Vision.* Oakland, Pacific Institute for Studies in Development, Environment and Security.

Government of India (1992). All India Report on Agricultural Census, 1985-1986. New Delhi, Government of India, Ministry of Agriculture.

Government of India (1996). Report of the Working Group on Minor Irrigation for Formulation of the Ninth Plan (1997-2002) Proposals. New Delhi, Government of India, Ministry of Water Resources.

Green, Donald E. (1973). *Land of the Underground Rain: Irrigation of the Texas High Plains, 1910-1970.* Austin: University of Texas Press, 1973.

Green, Donald E. (1992). "A History of Irrigation Technology Used to Exploit the Ogallala Aquifer", Chapter 2 of David E. Kromm and Stephen E. White, eds. *Groundwater Exploitation in the High Plains.* Lawrence: University Press of Kansas.

Grey, D. R. C., et al. (1995). Groundwater in the United Kingdom. A Strategic Study. Issues and Research Needs. Groundwater Forum Report FR/GF 1. Foundation for Water Research, Marlow, United Kingdom.

Gutentag, Edwin D., et al. (1984). Geohydrology of the High Plains Aquifer in Parts of Colorado, Kansas, Nebraska, New Mexico, Oklahoma, South Dakota, Texas and Wyoming, USGS Professional Paper 1400-B. Washington, D.C.: Government Printing Office.

Hardin, G. (1968). The Tragedy of the Commons. *Science* 162: 1243-1248.

Hayton, R. D., and A. Utton (1989). *Transboundary Groundwaters: The Bellagio Draft Treaty.* International Transboundary Resources Center. 414 pp.

Hazeldine, S., and D. Smythe (1997). Why was Sellafield rejected as a disposal site for radioactive waste? *Geoscientist* (7) 7. Geological Society, London.

Herbert, R., and B. Adams (1996). Groundwater Development in Unconsolidated Sedimentary Aquifers: A Developers Aid. British Geological Survey Report WC/96/06.

Herdt, R. W., and T. Wickham (1978). Exploring the gap between potential and actual rice yields: the Philippine case. *Economic Consequences of the New Rice Technology.* Los Banos, Philippines, International Rice Research Institute.

Ilich, N. (1996). Ganges River Basin Water Allocation Modelling Study. Calgary. Consultant Report to the World Bank.

Janakarajan, S. (1994). Trading In Groundwater: A Source of Power and Accumulation. In M. Moench (ed.), *Selling Water: Conceptual and Policy Debates over Groundwater Markets in India.* Ahmedabad, VIKSAT: 47-58.

Jones, M. J., (1985). The weathered zone aquifers of the basement complex areas of Africa. *Q.J.Eng. Geol.*, 18, 35-46.

Joshi, P. K., et al. (1995). Measuring Crop Damage due to Soil Salinity. *Strategic Change in Indian Irrigation.* M. S. A. Gulati. New Delhi, Macmillan India Limited.

Kahnert, F., and G. Levine (1989). *Key Findings, Recommendations, and Summary. Groundwater Irrigation and the Rural Poor: Options for Development in the Gangetic Basin*, Washington, D.C., the World Bank.

Kemper, K. (1996). The Cost of Free Water: Water Resource Allocation and Use in the Kuru Valley, Ceara, Northeast Brazil, Linkoping Studies in Arts and Sciences 137, Linkoping, Sweden, pp. 195-200.

Kingdom of the Netherlands (1992). India: Hydrological Investigations for the Santalpur and Sami-Harij RWSS, Kingdom of the Netherlands, Directorate General of International Cooperation; Government of India, Ministry of Agriculture; and Gujarat Water-Supply and Sewerage Board.

Koehler, C. (1995). Water Rights and the Public Trust Doctrine: Resolution of the Mono Lake Controversy. *Ecology Law Quarterly*, 22(3): 541-589.

Kromm, David E., and Stephen E. White (1983). Irrigator Response to Groundwater Depletion in Southwestern Kansas. *Environmental Professional* 5:106-115.

Kromm, David E., and Stephen E. White (1990). Water-Saving Practices by Irrigators in the High Plains, *Water Resources Bulletin*, vol. 26, pp. 999-1012.

Kromm, David E., and Stephen E. White (1991). Reliance on Sources of Information for Water-Saving Practices by Irrigators in the High Plains of the USA, *Journal of Rural Studies*, 7 (4) 411-421.

Kromm, David E., and Stephen E. White (1992a). The High Plains Ogallala Region. Chapter 1 of David E. Kromm and Stephen E. White, eds., *Groundwater Exploitation in the High Plains*, Lawrence: University Press of Kansas.

Kromm, David E., and Stephen E. White (1992b). Future Prospects. Chapter 11 of David E. Kromm and Stephen E. White, eds., *Groundwater Exploitation in the High Plains*, Lawrence: University Press of Kansas.

Lerner, D., et al. (1990). Groundwater Recharge. A Guide to Understanding and Estimating Natural Recharge. International Association of Hydrogeologists. *International Contributions to Hydrogeology.* Volume 8. Hannover, Heise, 345 pp.

Lerner, D. N., and N. R. G. Walton (1998). Contaminated Land and Groundwater: Future Directions. British Geological Society Engineering Geology Special Publication 14. Geological Society of London.

Lewis, W. J., et al. (1980). *The risk of groundwater pollution from on-site sanitation in developing countries.* IRCWD.

Livingstone, M. L. (1993). Normative and Positive Aspects of Institutional Economics: The Implications for Water Policy. *Water Resources Research* 29 (4), 815-821.

Lloyd, J. W., ed. (1981). *Case Studies in Groundwater Resources Evaluation.* Clarendon Press. Oxford.

MacFarlane, M. J. (1985). The Weathering Profile above Crystalline Basement Rocks under Tropical Weathering Conditions and in the Context of Hydrogeology. British Geological Survey. Report WD/OS/85/3. Keyworth, Nottingham.

Mähler, K. G. (1986). Comment on R. M. Solow, "On the Intergenerational Allocation of Natural Resources". *Scandinavian Journal of Economics*, 88: 151-152.

Mather, J. D., et al., eds. (1998). Groundwater Contaminants and their Migration. Geological Society Special Publication 128. Geological Society of London.

McFadden, (1989). Aspects of San Luis Valley Water in 1989: Administrative, Investigative, and Litigative. Water in the Valley: A 1989 Perspective on Water Supplies, Issues and Solutions in the San Luis Valley, Colorado. CGWA. Denver, Colorado Ground-Water Association: 111-123.

Meinzen-Dick, R. (1996). *Groundwater Markets in Pakistan: Participation and Productivity.* Washington, D.C., International Food Policy Research Institute.

Meinzer, O. E. (1920). Quantitative methods of estimating groundwater supplies. *Bulletin of the Geological Society of America* 31, pp. 329-328.

Metropolitan Water District (1995). IRP: Integrated Resources Plan Public Participation, Comprehensive Water Resource Management Strategies for Southern California. Document prepared by the Metropolitan Water District of Southern California, Los Angeles.

Miles, J. C., and P. D. Chambet (1995). Safe Yield of Aquifers. *Journal of Water Resources Planning and Management.* 121 (1).

Mitchell, B., ed. (1990). *Integrated Water Management: International Experiences and Perspectives.* Bellhaven Press. London. 225 pp.

Moench, M. (1991). *Social Issues in Western U.S. Groundwater Management: An Overview.* Oakland, Pacific Institute.

Moench, M. (1994a). Approaches to Groundwater Management: To Control or Enable. *Economic and Political Weekly* (September 24): A135-A146.

Moench, M., ed. (1994b). *Selling Water: Conceptual and Policy Debates over Groundwater Markets in India.* Ahmedabad, VIKSAT, Pacific Institute and Natural Heritage Institute.

Moench, M. (1995a). *Electricity Prices: A Tool for Groundwater Management in India.* Ahmedabad, VIKSAT.

Moench, M. (1995b). Allocating the Common Heritage: Debates over Groundwater Rights in India and the Western U.S. Paper presented at the first Open Meeting of the Human Dimensions of Global Environmental Change Community, Duke University, Durham, North Carolina.

Moench, M. (1996). *Groundwater Policy: Issues and Alternatives in India.* Colombo, International Irrigation Management Institute.

Moench, M. (1997). Local Water Management: Options and Opportunities in Yemen. Consultant Report to the World Bank. World Bank, Washington, D.C.

Moench, M., and H. Matzger (1994). Ground Water Availability for Drinking in Gujarat: Quantity, Quality and Health Dimensions. *Economic and Political Weekly* XXIX(13): A-31-A-14.

Moore, D., and D. Seckler (1985). *Water Scarcity in Developing Countries: Reconciling Development and Environmental Protection.* Proceedings of a round-table discussion. Washington, D.C., Environmental Defense Fund and Winrock International.

Morris, B. L., et al. (1994). The Impact of Urbanization on Groundwater Quality (project summary report). Keyworth, British Geological Survey.

Mundle, S., and M. G. Rao (1991). Volume and Composition of Government Subsidies in India, 1987-1988. *Economic and Political Weekly* (May 4): 1157-1172.

Murphy-Earl (1991). *Water and Water Rights.* vol 3., p. 59. The Mitchie Company. Charlottesville, Virginia

National Groundwater Association (1996). http://h20-ngwa.org

National Research Council (1992). *Water Transfers in the West: Efficiency, Equity and the Environment.* Washington, D.C., National Academy Press.

National Research Council (1997). *Valuing Groundwater: Economic Concepts and Approaches.* Committee on Valuing Groundwater, Washington, D.C., National Academy Press.

Norconsult (1982). Rukwa Water Master Plan, Volume 2. Water Development Atlas prepared for Norwegian Agency for International Development (NORAD). Oslo.

Ostrom, E. (1993). Design Principles in Long-Enduring Irrigation Institutions. *Water Resources Research* 29(7): 1907-1912.

Pague, C. A., and S. E. Simonson (1994). *Patterns of Rarity in the San Luis Valley of Colorado.* Fort Collins, Colorado Natural Heritage Program, Colorado State University.

Pedley, S., and G. Howard (1997). The public health implications of microbiological contamination of groundwater. *Quarterly Journal of Engineering Geology*, 30(2): 179-188.

Phadtare, P. N. (1988). *Geohydrology of Gujarat State.* Ahmedabad, Central Groundwater Board, West Central Region, pp. 103 and Appendices.

Pisharote, P. R. (1992). Rainfall Regime of Kutch District, Gujarat State, India. Workshop on Drought Management, Mandvi, Kutch.

Poffenberger, M., and B. McGean (1994). Policy Dialogue on Natural Forest Regeneration and Community Management, Asia Sustainable Forest Management Network. Research Network Report Number 5, East-West Center, Honolulu, Hawaii.

Porter, M. E., and C. van der Linde (1995). Green and Competitive: Ending the Stalemate. *Harvard Business Review*. September-October 1995. Cambridge, Massachusetts.

Postel, S. (1993). Water in Agriculture. P. Gleick (ed.). In *Water in Crisis.* New York, Oxford University Press: 56-66.

Pyne, R. G. D. (1995). *Groundwater Recharge and Wells. A Guide to Aquifer Storage and Recovery.* Lewis Publishers. Boca Raton. 376 pp.

Reisner, M. (1986). *Cadillac Desert.* New York, Viking Penguin.

Reisner, M., and S. Bates (1990). *Overtapped Oasis.* Washington, D.C., Island Press.

Repetto, R. (1994). The "Second India" Revisited: Population, Poverty, and Environmental Stress over Two Decades. Washington, D.C., World Resources Institute.

Ribot, J. C., et al., eds. (1996). *Climate Variability, Climate Change and Social Vulnerability in the Semi-Arid Tropics.* Cambridge, Cambridge University Press.

Robins, N. S., ed. (1998). *Groundwater Pollution, Aquifer Recharge and Vulnerability.* Geological Society Special Publication No. 130. The Geological Society, London.

Rogers, D. H., et al. (1997). Efficiencies and Water Losses of Irrigation Systems, MF-2243, Manhattan: Cooperative Extension Service, Kansas State University.

Rogers, P., et al. (1989). Eastern Waters Study: Strategies to Manage Flood and Drought in the Ganges-Brahmaputra Basin. Arlington, Irrigation Support Project for Asia and the Near East, USAID Project No. 3-7631510.

Rosegrant, M., and R. Schleyer (1994). *Tradable Water Rights: Experiences in Reforming Water Allocation Policy.* Arlington, Virginia.

Sadeque, S. Z. (1996). Nature's Bounty or Scarce Commodity—Competition and Consensus Over Ground Water Use in Rural Bangladesh. Annual Conference of the International Association for the Study of Common-Property, University of California, Berkeley.

Sauveplane, C. (1987). Analytical Modelling of Transient Flow to Wells in Complex Aquifer Systems. Ph.D. thesis, University of Alberta.

Shah, T. (1993). *Groundwater Markets and Irrigation Development: Political Economy and Practical Policy.* Bombay, Oxford University Press.

Shah, T., et al. (2000). The Global Groundwater Situation: Overview of Opportunities and Challenges. International Water Management Institute. Second World Water Forum 15.

Schiffler, M. (1998). *The Economics of Groundwater Management in Arid Countries.* GDI Book Series II. Frank Cass, London.

Seckler, D. U. (1996). *The New Era of Water Resource Managment: From Dry to Wet Water Savings.* International Irrigation Management Institute, Colombo.

Simmers, I., et al., eds. (1992). Selected papers on aquifer overexploitation. Puerto de la Cruz. Tenerife, Spain. April 15-19, 1991. International Association of Hydrogeologists, Heise.

Singh, C. (1991). *Water Rights and Principles of Water Resources Management.* Bombay, Indian Law Institute, N. M. Tripathi, PVT Ltd.

Smedly, P. L., et al. (1995). Vulnerability of shallow groundwater quality due to natural geochemical environments. British Geological Survey Technical Report WC/95/43.

Solanes, M. (1996). Institutional and Legal Issues Relevant to the Implementation of Water Markets. Report to the Secretary General. Committee on Natural Resources. Third Session. United Nations Economic and Social Council, E/C.7/1996/3.

Solanes, M., and F. G. Villareal (1999). A Comparative Assessment of Institutional and Legal Arrangements for Integrated Water Management. In G. J. Alaerts, et al., eds., *Water Sector Capacity Building: Concepts and Instruments.* Rotterdam: Balkema.

Solley, W. B., et al. (1998). Estimated Use of Water in the United States in 1995, U.S. Geological Survey Circular 1200, United States Government Printing Office.

Sophoceous, Marios, ed. (1998). Perspectives on Sustainable Development of Water Resources in Kansas. Lawrence: Bulletin 239. Lawrence: Kansas Geological Survey.

Sweeten, John M., and Wayne R. Jordan (1987). Irrigation Water Management for the Texas High Plains: A Research Summary. Texas Water Resources Institute. Technical Report No. 139. College Station: Texas A&M University.

Taylor, Johnathan G., et al. (1988). Adapting to Environmental Change: Perceptions and Farming Practices in the Ogallala Aquifer Region. In Emily E. Whitehead et al., ed., *Arid Lands: Today and Tomorrow*. Boulder: Westview Press, 665-684.

Templer, O. W. (1992). The Legal Context for Groundwater Use. Chapter 4 of David E. Kromm and Stephen E. White, eds., *Groundwater Exploitation in the High Plains*. Lawrence: University Press of Kansas.

Tsur, Y. (1990). The Stabilization Role of Groundwater when Surface-Water Supplies Are Uncertain: The Implications for Groundwater Development. *Water Resources Research*, 26(5): 811-818.

Tsur, Y. (1993). *The Economics of Conjunctive Ground and Surface-Water Irrigation Systems: Basic Principles and Empirical Evidence from Southern California 22.* Minneapolis, University of Minnesota.

United Nations (1960). *Large-Scale Groundwater Development.* Water Resources Development Centre. New York.

United Nations (1987). Groundwater Economics. Report of an International Symposium and Workshop Convened with the Government of Spain. Barcelona, Spain (19-23 October, 1987). United Nations Department for Technical Cooperation and Development, New York.

United Nations (1990). Schema Directeur de Mise en Valeur des Ressources en Eau du Mali. Volume 1. Rapport. United Nations Department for Technical Cooperation and Development. Project MLI/84/005. New York.

United Nations (1991). Interregional workshop on groundwater over-exploitation in developing countries. Gran Canaria, Spain. 20-24 April 1991. United Nations Department for Techical Cooperation and Development, New York.

United Nations (1995a). Review of United Nations Development Programme Supported Projects Executed by Department for Development Support and Management Services. New York. 168 pp.

United Nations (1995b). A review of the adverse effects of water resources development and use on groundwater and aquifers. Department for Development Support and Management Services. 70 pp. New York.

United Nations (1996a). *World Urbanization Prospects. The 1996 Revision.* United Nations Department of Economic and Social Affairs, Population Division. New York, 191 pp.

United Nations (1996b). Managing Water Resources for Large Cites and Towns. Report of the Habitat II International Conference, 18-21 March 1996, Beijing, China. United Nations Centre for Human Settlements, Nairobi.

United Nations (1996c). Indicators of Sustainable Development Methodology Sheets. Background Paper No. 15. United Nations Department for Policy Coordination and Sustainable Development.

United Nations (1998a). *World Population Prospects. The 1998 Revision.* Population Division, UNDESA, New York.

United Nations (1998b). Rapport d'Activités d'Appui au Processus de Planification Signer. Projet d'Appui au Plan Eau et Developpement NER/94/002. UNDESA, New York.

United Nations (1998c). Commission on Sustainable Development. Report of the Sixth Session. Economic and Social Council Supplement No. 9. New York.

United Nations (1999). Report of the expert group meeting on Strategic Approaches to Freshwater Management. Harare, Zimbabwe, 27-30 January. New York. Division for Sustainable Development, Department of Economic and Social Affairs, New York.

United States Bureau of the Census (1969, 1978, 1987, 1992). Census of Agriculture. U.S. Government Printing Office, Washington, D.C.

United States Geological Survey (1995). Water-Level Changes in the High Plains Aquifer, 1980 to 1994, Fact Sheet FS-215-95.

Vaidyanathan, A. (1994). *Second India Studies Revisited: Food, Agriculture and Water.* Madras, Madras Institute of Development Studies.

van der Merwe, B. (1998). Integrated Water Resource Management in Windhoek, Namibia. Report to UNCTAD meeting, Seminar on Strengthening Capacities in Developing Countries to Develop their Environmental Services Sector, Geneva, 20-22 July 1998.

VIKSAT/Pacific Institute (1993). *Proceedings of the Workshop on Water Management: India's Groundwater Challenge*, edited by M. Moench, S. Turnquist and D. Kumar, VIKSAT, Ahmedabad, Gujarat, December 14-16, pp. 199.

Waldron, J. (1994). The Edges of Life. Review of *Life's Dominion: An Argument about Abortion and Euthenasia* by Ronald Dworkin. London Review of Books. 12 May 1994. London.

White, Stephen E., and David E. Kromm (1995). Local Groundwater Management Effectiveness in the Colorado and Kansas Ogallala Region. *Natural Resources Journal*, vol. 35, pp. 275-307.

Wijdemans, R. T. J. (1995). Sustainability of Groundwater for Water-Supply: Competition Between the Needs for Agriculture and Drinking Water. In M. Moench (ed.), *Groundwater Availability and Pollution: The Growing Debate over Resource Condition in India*. Ahmedabad, VIKSAT. pp. 160.

Wilson, L. G., et al. (1998). Hydrogeologic uncertainties and policy implications: The Water Consumer Protection Act of Tucson, Arizona. *Hydrogeology Journal*, 6:3-14.

Winpenny, J. (1994). *Managing Water as an Economic Resource*. Routledge, London.

World Bank (1983). Sudan Livestock Route Company. Staff Appraisal Report. International Bank for Reconstruction and Development, Washington, D.C.

World Bank (1993). *Water Resources Management*. International Bank for Reconstruction and Development, Washington, D.C.

World Bank (1994). Pakistan Irrigation and Drainage: Issues and Options Report No. 11884-PAK, March 25, 1994. International Bank for Reconstruction and Development, Washington, D.C.

World Bank (1996). An Evaluation of the UNDP–World Bank Water Supply and Sanitation Programme. Report of an Independent Team. 106 pp.

World Bank (1998a). *Groundwater Regulation and Management Report*, India—Water Resources Management Sector Review. International Bank for Reconstruction and Development, Washington, D.C.

World Bank (1998b). Groundwater in Urban Development: Assessing Management Needs and Formulating Policy Strategies. World Bank Technical Paper No. 390.

World Commission on Environment and Development (1987). *Our Common Future*. Oxford University Press, London.

Water Resources Assessment of Yemen (1995). The Water Resources of Yemen. Sana'a. Report 35. General Department of Hydrology, Republic of Yemen, TNO Institute of Applied Geoscience, and Kingdom of the Netherlands.

Wright E. P., and W. Burgess, eds. (1992). *The Hydrogeology of Crystalline Basement Aquifers in Africa*. Geological Society of London Special Publication, No. 66, London 264 pp.